SpringerBriefs in Optimization

Series Editors

Panos M. Pardalos
János D. Pintér
Stephen M. Robinson
Tamás Terlaky
My T. Thai

SpringerBriefs in Optimization showcases algorithmic and theoretical techniques, case studies, and applications within the broad-based field of optimization. Manuscripts related to the ever-growing applications of optimization in applied mathematics, engineering, medicine, economics, and other applied sciences are encouraged.

For further volumes:
http://www.springer.com/series/8918

Adam B. Levy

Stationarity and Convergence in Reduce-or-Retreat Minimization

 Springer

Adam B. Levy
Department of Mathematics
Bowdoin College
Brunswick, ME
USA

ISSN 2190-8354   ISSN 2191-575X (electronic)
ISBN 978-1-4614-4641-5   ISBN 978-1-4614-4642-2 (eBook)
DOI 10.1007/978-1-4614-4642-2
Springer New York Heidelberg Dordrecht London

Library of Congress Control Number: 2012943990

Mathematics Subject Classification (2010): 90C99, 49M99, 65K05, 90-02, 49-02

Printed on acid-free paper

Springer is part of Springer Science+Business Media (www.springer.com)

To Sarah, Jonah, and Emma.

Preface

This monograph presents and analyzes a unifying framework for a wide variety of numerical minimization methods. Our "reduce-or-retreat" framework is a conceptual method outline that covers essentially any method whose iterations choose between the two options of reducing the objective in some way at a trial point, or (if reduction is not possible) retreating to a closer set of trial points. Included in this category are many derivative-based methods (which depend on the objective gradient to generate trial points), as well as many derivative-free and direct methods that generate trial points either from some model of the objective or from some designated pattern in the variable space. These different types of methods are usually considered entirely separately, but our framework reveals essential similarities and is partly inspired by the visual similarity illustrated in Fig. 1.

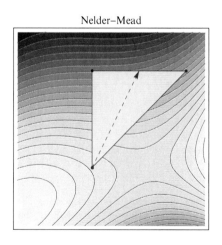

Fig. 1 Visual similarity between methods

At each step of the direct method on the left (Nelder–Mead), an improved point is sought in the direction away from the highest-valued vertex and toward the two lower-valued vertices of a triangle. Given only the values at the vertices, this direction is a practical facsimile of the descent direction along which an improved point is sought in the derivative-based method on the right (steepest descent).

By aligning derivative-free and direct methods within the same framework as derivative-based methods, we hope to do many things; one of which is to encourage the construction of new methods. We illustrate this process here by developing two generalizations of classic derivative-based methods (a line-search and a trust-region method), which accommodate non-smooth objectives. We also hope that our framework will inspire new theoretical developments as companions to results from across the traditional divides. We illustrate how this can work through a detailed analysis of our two generalized derivative-based methods, alongside similar analyses of two direct methods (a general pattern-search method and the classic Nelder–Mead method).

In addition to providing a bridge for theory through our framework, we extend and broaden the traditional convergence analysis in several important ways. Until now, this analysis has been carried out independently and in specialized ways for each of the particular methods within our framework. Even so, the influence of traditional derivative-based methods is seen in the typical analysis of the other methods, where it remains a focus to show that iterates x^k "approach stationarity" by generating gradients that approach zero. This kind of gradient-based property is necessarily designed for smooth objectives $f \in \mathscr{C}^1$, even though derivative-free and direct methods are constructed specifically to accommodate non-smooth objectives. We extend the notion of approaching stationarity to non-smooth objectives, which allows us to more completely analyze the convergence of all of the methods within the framework.

As part of our convergence analysis, we explore the crucial role of "non-degeneracy" in deducing an approach to stationarity. Existing reduce-or-retreat methods are sometimes structured to automatically ensure non-degeneracy, and our framework easily accommodates this approach. However, we also illustrate a less typical approach to ensuring non-degeneracy through an optional (and occasional) degeneracy check that resets some elements of the method to avoid degeneracy. Finally, we embrace here the practical alternative of checking the output after a particular run of a method to see if an appropriate non-degeneracy condition is satisfied (in which case, the desired convergence result can be deduced). The more traditional approaches allow one to say that *any* particular run of the method approaches stationarity, but this universality comes with a price of additional structure in the method (and the associated computational burdens). In contrast, the practical alternative of an a posteriori check can give us confidence in the output we obtain from a particular run of a method, without having to somehow embed non-degeneracy conditions in the method itself. This alternative is particularly attractive for methods (like Nelder–Mead) that are generally effective in practice, but lack a robust global theory.

Another way in which we broaden the traditional analysis is by identifying seven different "scenarios" that cover the possibilities for any run of a method within our reduce-or-retreat framework and then analyzing the convergence properties of the different elements computed at each iteration (including "iterate-sets" whose existing convergence analysis is very sparse). For example, one "scenario" we analyze is an (eventual) uninterrupted sequence of retreat iterations. The effect of this on the elements computed at each iteration is very different from that of the scenario when there is instead an (eventual) uninterrupted sequence of reductions. A systematic analysis of this kind is essentially unprecedented, though it is motivated by two examples from the Nelder–Mead literature: Lagarias et al. [5, Lemma 3.5] proved that the scenario of all reductions occurs in Nelder–Mead whenever the objective is strictly convex, and Mckinnon [6] presented an example of an objective function for which a particular run of the Nelder–Mead method produced an uninterrupted sequence of one type of reduction. The latter example was in fact designed to show that bad convergence behavior can occur in Nelder–Mead even with a seemingly well-behaved (smooth, strictly convex) objective function. Interestingly, this example fails an a posteriori check of the corresponding non-degeneracy condition, and so would be identified as an unreliable run by our practical approach.

Some quite general methods for minimization have been explored before in various contexts (e.g., in [3, 4]), but the accompanying analyses have focussed almost exclusively on traditional notions of convergence. Our framework overlaps these methods, and our convergence analysis complements and expands the existing results.

Harpswell, ME, USA   Adam B. Levy

Contents

1	**A Framework for Reduce-or-Retreat Minimization**		1
	1.1	Iteration-Generated Elements	1
	1.2	Reduction, Retreat, and Termination	2
	1.3	Framework Outline	3
	1.4	Descent	4
	1.5	Lower-Diminishing Target-Gaps Γ^k	5
	1.6	Approaching Stationarity	6
		1.6.1 Lipschitz ∂f	8
		1.6.2 Full-Sequence Stationarity	9
	1.7	Stationarity-Inducing Target-Gaps	10
2	**Particular Reduce-or-Retreat Methods**		13
	2.1	Generalized Line-Search Method	13
		2.1.1 Method Outline	14
		2.1.2 Descent	15
		2.1.3 Lower-Diminishing Target-Gaps	15
		2.1.4 Approaching Stationarity	17
	2.2	Generalized Trust-Region Method	18
		2.2.1 Method Outline	21
		2.2.2 Descent	23
		2.2.3 Lower-Diminishing Target-Gaps	23
		2.2.4 Approaching Stationarity	27
	2.3	General Pattern-Search Method	28
		2.3.1 Method Outline	28
		2.3.2 Descent	29
		2.3.3 Lower-Diminishing Target-Gaps	30
		2.3.4 Approaching Stationarity	30
	2.4	Nelder–Mead Method	33
		2.4.1 Method Outline	34
		2.4.2 Lower-Diminishing Target-Gaps	36
		2.4.3 Approaching Stationarity	37

2.5   Incorporating a Degeneracy Check 40
       2.5.1   Non-degeneracy Conditions 40

3 Scenario Analysis 43
3.1   Convergence of Trial-Sizes Δ^k 45
3.2   Convergence of Iterate-Sets X^k 46
       3.2.1   Iterate-Diameter and Iterate-Radius 47
       3.2.2 f-Stability 49

References 55

Chapter 1
A Framework for Reduce-or-Retreat Minimization

We introduce our framework, including all of the elements generated at each iteration, as well as the major components of reduction and retreat. The property of nonincreasing function values inherent in the framework is strengthened via the definition of a descent method, and lower-diminishing target-gaps are explored as the result of a successful run of a method within the framework. Finally, we develop our extended notion of approaching stationarity and link this property both to stationary points and to lower-diminishing target-gaps.

Our framework is a structured iterative process that generates three primary elements at each iteration. Depending on the conditions met by these elements, the iterative process is terminated or updated according to two different schemes (reduction and retreat).

1.1 Iteration-Generated Elements

Each iteration (indexed by k) of a reduce-or-retreat method generates an *iterate-set* $X^k \subseteq \mathbb{R}^n$ consisting of a finite number m of n-vectors, as well as a *trial-size* $\Delta^k \geq 0$. Depending on the particular method, the trial-size Δ^k could represent a step-size, a trust-region radius, or even the volume of the iterate-set. The *iterate* x^k is an element of the iterate-set X^k achieving lowest f-value:

$$x^k \in \operatorname*{argmin}_{x \in X^k} f(x).$$

Some well-known methods within our framework use a singleton iterate-set which thus coincides with the iterate $X^k = \{x^k\}$. The trial-size Δ^k and iterate-set X^k are used to compute a nonempty *trial-set* $T^k \subseteq \mathbb{R}^n$ of d candidate n-vectors to replace an element of the current iterate-set; and the trial-size Δ^k gets its name from its role in this process.

A.B. Levy, *Stationarity and Convergence in Reduce-or-Retreat Minimization*, SpringerBriefs in Optimization, DOI 10.1007/978-1-4614-4642-2_1, © Adam B. Levy 2012

1.2   Reduction, Retreat, and Termination

The main step in the framework involves either a reduction or a retreat. We first choose an "outgoing" point $x_{\text{out}}^k \in X^k$ which might be removed from the iterate-set and then attempt to identify a satisfactory trial point $\xi^k \in T^k$ via a *reduction test*. If such a trial point ξ^k can be found, it replaces x_{out}^k in the iterate-set X^k. On the other hand, if there is not a satisfactory trial point, then we "retreat" by shrinking the trial-size $\Delta^{k+1} \in [\theta_L \Delta^k, \theta_U \Delta^k]$ for some fixed fractions $\theta_L \le \theta_U$ in $(0, 1)$. Also, in this case, the iterate-set X^{k+1} is updated according to some rule specified by the particular reduce-or-retreat method, always in such a way that the new iterate x^{k+1} does not have a higher f-value than the old iterate x^k:

$$f(x^{k+1}) = \min_{x \in X^{k+1}} f(x) \le \min_{x \in X^k} f(x) = f(x^k). \tag{1.1}$$

No change $X^{k+1} = X^k$ is one relatively common such update rule.

Different particular methods within the reduce-or-retreat framework also use different rules to choose an outgoing point $x_{\text{out}}^k \in X^k$, and to define a "satisfactory" trial point. However, in every case, the satisfactory trial point $\xi^k \in T^k$ will have a lower f-value than the outgoing point x_{out}^k that it replaces in the iterate set. This guarantees a reduction of the cumulative f-values of the iterate-set:

$$\sum_{x \in X^{k+1}} f(x) = \sum_{x \in \{\xi^k\} \cup X^k \setminus \{x_{\text{out}}^k\}} f(x)$$

$$= f(\xi^k) + \left(\sum_{x \in X^k} f(x) \right) - f(x_{\text{out}}^k)$$

$$< \sum_{x \in X^k} f(x),$$

which motivates the term "reduction."

The stopping criterion is $\Gamma^k = 0$, in terms of *target-gaps* $\Gamma^k \ge 0$ evaluated at each iteration. Particular reduce-or-retreat methods are typically designed to drive a corresponding target-gap Γ^k to zero, since a small target-gap indicates that the intended "target" of the method has been nearly reached. Our framework allows very general target-gaps Γ^k, but we will be especially interested in relationships between particular choices of target-gaps Γ^k and the set of "subgradients" $\partial f(x^k)$ of f at x^k. Our interest in subgradients is motivated by the fact that the inclusion $0 \in \partial f(x^k)$ identifies the iterate x^k as a *stationary point* of f, which is an important necessary condition for x^k to be a local minimizer [7, Theorem 10.1]. For smooth objective functions $f \in \mathscr{C}^1$, the set of subgradients reduces to the gradient $\partial f(x^k) = \{\nabla f(x^k)\}$ [7, Exercise 8.8], and the stationary-point inclusion $0 \in \partial f(x^k)$ reduces to the familiar Fermat rule $0 = \nabla f(x^k)$.

Some existing reduce-or-retreat methods within our framework do not explicitly include a stopping criterion of any kind, and others essentially use very particular target-gaps. The structure of our framework encourages innovations in both cases since the target-gaps may be defined in different ways.

1.3 Framework Outline

Reduce-or-Retreat Minimization Framework

Step 1. Initialize the iterate-set X^0 and the trial-size $\Delta^0 > 0$, and generate the corresponding trial-set T^0. Fix $\theta_L \leq \theta_U$ in $(0,1)$ and $\theta_{\text{red}} \in (0,\infty)$, and set $k = 0$.

Step 2. Stopping criterion.

- (Terminate) If the target-gap Γ^k is zero, then:

  - Set $\bar{x} = x^k$ as the identified target
  - Fix all later iterate-sets, trial-sizes, and target-gaps:

  $$X^{i+1} = \{\underbrace{\bar{x}, \bar{x}, \ldots, \bar{x}}_{m \text{ copies}}\}, \Delta^{i+1} = 0, \text{and} \Gamma^{i+1} = 0 \quad \text{for } i \geq k.$$

- Otherwise, continue to Step 3.

Step 3. Reduce or retreat. Choose an outgoing point $x^k_{\text{out}} \in X^k$.

- (Reduce) If the reduction test identifies a satisfactory trial point $\xi^k \in T^k$, then:

  - Swap ξ^k for x_{out} in the iterate-set:

  $$X^{k+1} = \left\{\xi^k\right\} \cup X^k \setminus \left\{x^k_{\text{out}}\right\}.$$

  - Choose $\Delta^{k+1} \geq \theta_{\text{red}} \Delta^k$.
  - Then continue to Step 4.

- (Retreat) Otherwise:

  - Update the iterate-set X^{k+1} without increasing the f-value of the iterate (1.1).
  - Shrink the trial-size $\Delta^{k+1} \in [\theta_L \Delta^k, \theta_U \Delta^k]$.
  - Then continue to Step 4.

Step 4. Compute the new target-gap Γ^{k+1}, and generate the new trial-set T^{k+1}; increase the iteration-index by 1, and return to Step 2.

1.4   Descent

Recall that the satisfactory trial point $\xi^k \in T^k$ will have a lower f-value than the outgoing point x_{out}^k that it replaces in the iterate set, and recall that this guarantees a reduction of the cumulative f-values of the iterate-set. Notice that this also guarantees that the f-value of the iterate $x^k \in \operatorname*{argmin}_{x \in X^k} f(x)$ will never increase after a reduction step:

$$
\begin{aligned}
f\left(x^{k+1}\right) = \min_{x \in X^{k+1}} f(x) &= \min_{x \in \{\xi^k\} \cup X^k \setminus \{x_{\text{out}}^k\}} f(x) \\
&= \min\left\{ f(\xi^k), \min_{x \in X^k} f(x) \right\} \\
&\leq \min_{x \in X^k} f(x) = f(x^k).
\end{aligned}
\tag{1.2}
$$

Since we already know that the f-value of the iterate does not increase after a retreat step, we can conclude that the iterates have at least a convergent subsequence when the initial level set is bounded.

Proposition 1. *If the initial level set*

$$
\operatorname*{lev}_{f(x^0)} f := \{x \in \mathbb{R}^n | f(x) \leq f(x^0)\}
$$

is bounded, then the sequence of iterates x^k is bounded and has a convergent subsequence.

Proof. From the bounds (1.1) and (1.2) ensuring that the f-value of the iterate is never increased after retreat or reduction, we conclude that $f(x^k) \leq f(x^0)$ for all iterations. Thus, we know that every iterate x^k is in the bounded set $\operatorname*{lev}_{f(x^0)} f$, and there is a convergent subsequence as claimed.   □

For the f-value of the iterate x^k to actually decrease after reduction (and not simply avoid increase), the incoming trial point ξ^k needs to have a lower f-value than every point in the iterate-set X^k (and not just the outgoing point x_{out}^k). This property has traditional importance, so we say that a sequence of iterates x^k generated by a reduce-or-retreat method is *descending* if the iterates satisfy $f(x^{k+1}) < f(x^k)$ after all iterations inducing a reduction. A reduce-or-retreat method is called a *descent method* if it always generates descending sequences of iterates x^k.

Note that the term "descent method" is sometimes used to indicate further that if the iterate is not a stationary point, then a reduction will be induced after the trial-size becomes small enough. We address this kind of additional property much more generally in the following section.

1.5 Lower-Diminishing Target-Gaps Γ^k

When we first introduced target-gaps Γ^k, we mentioned that reduce-or-retreat methods are intended to diminish them. In this section, we will carefully explore conditions under which at least a subsequence of target-gaps Γ^k is certain to diminish. For simplicity, we use the terminology *lower-diminishing* to signify when at least a subsequence of values approaches zero, and we use the partial arrow "$\rightharpoonup$" as a compact way of indicating the lower limit encoding this property. Thus, we are interested in exploring when

$$\liminf_{k\to\infty} \Gamma^k = 0, \quad \text{equivalently} \quad \Gamma^k \rightharpoonup 0.$$

The key to our results will be establishing relationships between the lower limits of target-gaps and other sequences of scalars (which themselves may then be directly shown to be lower-diminishing).

In order to study such relationships in general, we say that the scalars γ^k are *lower-diminishing with* another sequence of scalars δ^k if

$$\delta^k \rightharpoonup 0 \quad \Rightarrow \quad \gamma^k \rightharpoonup 0.$$

We now show that this property is equivalent to the existence of an iteration-index K such that the following condition holds:

$$\gamma^k \geq \gamma > 0 \quad \forall k \geq K \quad \Rightarrow \exists \delta > 0 \quad \text{such that} \quad \delta^k \geq \delta \quad \forall k \geq K. \tag{1.3}$$

Lemma 1. *There exists an iteration-index K such that condition (1.3) holds if and only if the γ^k are lower-diminishing with δ^k.*

Proof. ($\Rightarrow$) If the δ^k are lower-diminishing, there can be no lower-bound $\delta > 0$ for which the right side of the implication in Eq. (1.3) holds. According to the condition (1.3) then, there can be no lower-bound $\gamma > 0$ for which $\gamma^k \geq \gamma$ for all $k \geq K$. We conclude that the γ^k are also lower-diminishing in this case.

($\Leftarrow$) There are two possibilities to consider. First, if the δ^k are not lower-diminishing, then the right side of the implication in Eq. (1.3) holds for some iteration-index K regardless of the convergence behavior of γ^k. Second, if the δ^k are lower-diminishing, then by assumption we conclude that the γ^k are lower-diminishing. In this case, we conclude that for any iteration-index K, there will not be a lower-bound $\gamma > 0$ triggering the condition on the left side of the implication in Eq. (1.3); so condition (1.3) holds vacuously. □

Because of how trial-sizes Δ^k are updated in the reduce-or-retreat framework, a related condition to Eq. (1.3) ensures that the target-gaps Γ^k are lower-diminishing with the trial-sizes.

Proposition 2. *If the condition*

$$\Gamma^k \geq \gamma > 0 \quad \forall k \quad \Rightarrow \exists \Delta > 0 \text{ such that } \Delta^{k+1} \geq \Delta^k \text{ whenever } \Delta^k \leq \Delta \quad (1.4)$$

holds, then the target-gaps Γ^k are lower-diminishing with the trial-sizes Δ^k.

Proof. According to Lemma 1, we simply need to show that the condition

$$\Delta^{k+1} \geq \Delta^k \text{ whenever } \Delta^k \leq \Delta \qquad (1.5)$$

implies a lower-bound $\delta > 0$ on the trial-sizes Δ^k, whenever $\Gamma^k \geq \gamma > 0$ for all k. Assuming $\Gamma^k \geq \gamma > 0$ for all k, we know that termination does not occur and that the trial-sizes Δ^k are all strictly greater than zero (recall that $\Delta^0 > 0$). Since reduction and retreat at most shrink the trial-size by factors of θ_{red} and θ_L, respectively, condition (1.5) implies that the trial-sizes can never fall below $\delta := \min\{\theta_{\text{red}}\Delta, \theta_L\Delta, \Delta^0\}$.   □

Knowing that the target-gaps Γ^k are lower-diminishing is especially useful if we can use that knowledge to deduce something about the minimization of f. In the next section we focus on stationarity as exactly the kind of property we would like to be able to deduce.

1.6   Approaching Stationarity

A traditional focus in the convergence analyses of minimization methods for smooth objective functions $f \in \mathscr{C}^1$ is on conditions ensuring that at least a subsequence of iterates has gradients approaching the zero vector:

$$\liminf_{k \to \infty} \|\nabla f(x^k)\| = 0, \quad \text{equivalently} \quad \|\nabla f(x^k)\| \to 0.$$

This analysis can be extended to non-smooth objective functions via "subgradients" (also, "Mordukhovich's subgradients"; see [7, p. 346]). Recall from [7, Definition 8.3] that v is a *subgradient of f at $\bar{x}$*, denoted $v \in \partial f(\bar{x})$, if and only if there are sequences of points $x^k \to \bar{x}$ and "regular subgradients" $v^k \to v$ with $f(x^k) \to f(\bar{x})$. The *regular subgradients $v^k \in \hat{\partial} f(x^k)$* are the vectors satisfying

$$\liminf_{x \to x^k, x \neq x^k} \frac{f(x) - f(x^k) - \langle v^k, x - x^k \rangle}{\|x - x^k\|} \geq 0.$$

When the objective function is smooth $f \in \mathscr{C}^1$, both kinds of subgradient sets reduce to the gradient [7, Exercise 8.8].

A lower-semicontinuous function f with a subgradient at x is called *regular* at x if the two kinds of subgradient sets coincide $\hat{\partial} f(x) = \partial f(x)$ (see [7, Corollary 8.11]), and the class of regular functions is quite broad. For instance, it includes convex functions [7, Example 7.27] and *lower-$\mathscr{C}^1$* functions (max functions constructed from collections of $\mathscr{C}^1$ functions [7, Theorem 10.31]). In general, subgradients enjoy a more robust calculus than regular subgradients (see [7, p. 301]), so we use subgradients to define our notion of stationarity.

We say that the subsequence of iterates *approaches stationarity* if there is a lower-diminishing sequence of smallest-norm subgradients:

$$\liminf_{k\to\infty} \inf_{v\in\partial f(x^k)} \|v\| = 0, \quad \text{equivalently} \quad \inf_{v\in\partial f(x^k)} \|v\| \to 0.$$

Notice that this property does not necessarily imply that a subsequence of iterates x^k converges to a stationary point $\bar{x} \in (\partial f)^{-1}(0)$. To ensure the latter property, we rely on the "outer semi-continuity" of the subgradient mapping. Following [7, Definition 5.4], we say that a set-valued mapping $S: \mathbb{R}^n \rightrightarrows \mathbb{R}^d$ is *outer semicontinuous at $\bar{x}$* if the inclusion holds that

$$\left\{v \mid \exists x^k \to \bar{x},\ \exists v^k \to v \text{ with } v^k \in S(x^k)\right\} \subseteq S(\bar{x}).$$

Rockafellar and Wets [7, Proposition 8.7] shows that for any point $\bar{x}$ where f is finite, the subgradient mapping $S(x) = \partial f(x)$ is outer semicontinuous at $\bar{x}$ *with respect to f-attentive convergence*, which means the inclusion holds that

$$\left\{v \mid \exists (x^k, v^k) \to (\bar{x}, v) \text{ with } f(x^k) \to f(\bar{x}) \text{ and } v^k \in \partial f(x^k)\right\} \subseteq \partial f(\bar{x}). \qquad (1.6)$$

The following result shows when an approach to stationarity can guarantee that the iterates actually approach a stationary point.

Theorem 1. *Suppose f is continuous and the iterates x^k are bounded. If a subsequence of iterates approaches stationarity, then a subsequence of iterates converges to a stationary point $\bar{x} \in (\partial f)^{-1}(0)$:*

$$\inf_{v\in\partial f(x^k)} \|v\| \to 0 \quad \Longrightarrow \quad \inf_{\bar{x}\in(\partial f)^{-1}(0)} \|x^k - \bar{x}\| \to 0. \qquad (1.7)$$

Proof. By assumption, there is a subsequence of iterations with $v^k \in \partial f(x^k)$ and $v^k \to 0$. Since the iterates are bounded, we know there is a sub-subsequence with $x^k \to \bar{x}$ for some limit point $\bar{x} \in \mathbb{R}^n$. Since f is continuous, we know $f(x^k) \to f(\bar{x}) < \infty$ for that sub-subsequence, so the outer semicontinuity inclusion (1.6) guaranteed by [7, Proposition 8.7] gives $0 \in \partial f(\bar{x})$. $\qquad\square$

Remark 1. The result of Theorem 1 holds even when f is not continuous, as long as $f(\bar{x}) < \infty$ and the sub-subsequence of iterates $x^k \to \bar{x}$ obeys *f-attentive convergence* $f(x^k) \to f(\bar{x})$.

Remark 2. If the iterates x^k happen to converge, then we can apply Theorem 1 to conclude that the limit is a stationary point.

Remark 3. The implication (1.7) has practical importance since the "approach to stationarity" defined by the subgradient-norms on the left side of Eq. (1.7) is something which we can assess directly by computing subgradients or by testing conditions we will develop in the following sections. On the other hand, the implied (and desirable) result that a subsequence of iterates converges to a stationary point is harder to confirm directly since zero-norm subgradients often compute as very-small-norm subgradients.

Notice that the opposite implication to Eq. (1.7) is not necessarily automatic. For instance, consider the objective function $f(x) = |x|$ taking the absolute value of a single variable, and the (bounded) sequence of iterates $x^k := \frac{1}{k}$. This sequence of iterates converges to the stationary point $\bar{x} = 0 \in (\partial f)^{-1}(0)$; however, the corresponding subgradients all satisfy $\partial f(x^k) = \{1\}$; so there is no approach to stationarity. To ensure the opposite implication to Eq. (1.7), we need additional assumptions which we explore now.

1.6.1   Lipschitz ∂f

Following [7, Definition 9.26], we say that a set-valued mapping $S : \mathbb{R}^n \rightrightarrows \mathbb{R}^d$ is *Lipschitz continuous on* $X \subseteq \mathbb{R}^n$ if there exists a constant $L > 0$ such that

$$S(x') \subseteq S(x) + L \|x' - x\| \quad \text{for all } x, x' \in X. \tag{1.8}$$

We want to apply this property to the subgradient mapping $S(x) = \partial f(x)$. Notice that when $f \in \mathscr{C}^{1,1}$, the subgradient mapping $S(x) = \partial f(x) = \{\nabla f(x)\}$ is Lipschitz continuous on any neighborhood $X \subseteq \mathbb{R}^n$ of $\bar{x} \in (\partial f)^{-1}(0)$. The next result reverses the implication in Theorem 1.

Theorem 2. *If a subsequence of iterates x^k converges to a stationary point $\bar{x} \in (\partial f)^{-1}(0)$, and if the subgradient mapping $S(x) = \partial f(x)$ is Lipschitz continuous on a neighborhood $X \subseteq \mathbb{R}^n$ of $\bar{x}$, then a subsequence of iterates approaches stationarity.*

Proof. We apply the Lipschitz continuity assumption with $x' := \bar{x}$ and $x := x^k$ for the subsequence of iterates x^k converging to $\bar{x}$. Since $0 \in \partial f(\bar{x}) = S(x')$ in this case, it follows from Eq. (1.8) that there is a $v^k \in \partial f(x^k) = S(x)$ satisfying $\|v^k - 0\| \leq L \|\bar{x} - x^k\|$. The result follows since $x^k \to \bar{x}$ for this subsequence.   □

Remark 4. For convex objective functions f, Lipschitz continuity of the subgradient mapping as in Theorem 2 is actually quite a restrictive assumption, and essentially demands that f is smooth $\big($see [7, p. 572]$\big)$.

1.6.2 Full-Sequence Stationarity

It is of course a better result to deduce an approach to stationarity for the full sequence of iterates (and not just a subsequence). In the traditional convergence analysis of reduce-or-retreat methods when $f \in \mathscr{C}^1$, knowing that the sequence of iterates converges allows us to make this extension.

Proposition 3. *For $f \in \mathscr{C}^1$, if the iterates converge $x^k \to \bar{x}$ and a subsequence approaches stationarity $\|\nabla f(x^k)\| \to 0$, then the entire sequence of iterates approaches stationarity $\|\nabla f(x^k)\| \to 0$. Moreover, the limit point $\bar{x}$ of the sequence of iterates is a stationary point $0 = \nabla f(\bar{x})$.*

Proof. By assumption, there is a subsequence with $\|\nabla f(x^k)\| \to 0$, so the convergence of the iterates $x^k \to \bar{x}$ together with the continuity of ∇f means that the entire sequence of $\|\nabla f(x^k)\|$ follows the subsequence.

The final statement that $0 = \nabla f(\bar{x})$ follows from Theorem 1. $\qquad \square$

We can generalize Proposition 3 to non-smooth objective functions by appealing to another generalized continuity property. Following [7, Definition 5.4], we say that the subgradient mapping is *inner semicontinuous at $\bar{x}$ with respect to f-attentive convergence* if

$$\partial f(\bar{x}) \subseteq \left\{ v \mid \forall x^k \to \bar{x} \text{ with } f(x^k) \to f(\bar{x}), \exists v^k \to v \text{ with } v^k \in \partial f(x^k) \right\}. \quad (1.9)$$

Proposition 4. *For any continuous f whose subgradient mapping is inner semicontinuous at $\bar{x}$ with respect to f-attentive convergence (1.9), if the iterates converge $x^k \to \bar{x}$ and a subsequence approaches stationarity*

$$\inf_{v \in \partial f(x^k)} \|v\| \to 0,$$

then the entire sequence of iterates approaches stationarity

$$\inf_{v \in \partial f(x^k)} \|v\| \to 0,$$

and the limit point $\bar{x}$ of the sequence of iterates is a stationary point $0 \in \partial f(\bar{x})$.

Proof. First, we appeal to Theorem 1 to conclude that $0 \in \partial f(\bar{x})$. Then the convergence of the iterates $x^k \to \bar{x}$, the continuity of f, and the inner semicontinuity inclusion (1.9) together ensure that there is a corresponding sequence of subgradients $v^k \in \partial f(x^k)$ with $v^k \to 0$. The result follows immediately from this. $\qquad \square$

Remark 5. The result of Proposition 4 holds even when f is not necessarily continuous, as long as the sequence of iterates $x^k \to \bar{x}$ obeys f-attentive convergence $f(x) \to f(\bar{x})$.

1.7   Stationarity-Inducing Target-Gaps

We have already analyzed lower-diminishing target-gaps Γ^k, and we now link this to the property of a subsequence of the iterates x^k approaching stationarity. We say that the target-gaps Γ^k are *stationarity-inducing* if the smallest-norm subgradients are lower-diminishing with Γ^k:

$$\Gamma^k \to 0 \quad \Rightarrow \quad \inf_{v \in \partial f(x^k)} \|v\| \to 0. \tag{1.10}$$

Trivially, the target-gaps $\Gamma^k := \nabla f(x^k)$ are stationarity-inducing, as are the non-smooth variants defined by $\Gamma^k := \inf_{v \in \partial f(x^k)} \|v\|$.

We will see in the sequel that having *nondegenerate* trial-sets T^k is crucial for deducing when the target-gaps are stationarity-inducing, where "degeneracy" will often be defined via a certain vector making an obtuse angle with all of the vectors in another set. To streamline these discussions, we define the (smallest) *angle between* the vectors $a \in \mathbb{R}^n$ and $b \in \mathbb{R}^n$:

$$\theta(a,b) := \arccos\left(\frac{\langle a, b\rangle}{\|a\|\,\|b\|}\right),$$

with the convention that 0 is the smallest angle between any vector and the zero-vector: $\theta(a, 0) = \theta(0, b) = 0$. We use this to define the *angular distance* from a vector $a \in \mathbb{R}^n$ to a set $B \subseteq \mathbb{R}^n$

$$\mathrm{angdist}(a, B) := \inf_{b \in B} \theta(a,b),$$

as well as the companion notion of *cosine-distance* from $a \in \mathbb{R}^n$ to $B \subseteq \mathbb{R}^n$

$$\mathrm{cosdist}(a, B) := \cos\left(\mathrm{angdist}(a, B)\right) = \sup_{b \in B} \frac{\langle a, b\rangle}{\|a\|\,\|b\|}. \tag{1.11}$$

When the set B is empty, the cosine-distance evaluates to negative infinity: $\mathrm{cosdist}\,(a, \emptyset) = -\infty$.

We will use cosine-distance to define non-degeneracy via bounds like cosdist $(a, B) \geq \varepsilon > 0$, which ensures that the vector a makes an acute angle with some vector in B. For example, when the objective functions are smooth $f \in \mathscr{C}^1$, and the trial-sets T^k are defined by search directions d^k away from the iterates x^k, non-degeneracy amounts to

$$\mathrm{cosdist}\left(d^k, \left\{-\nabla f(x^k)\right\}\right) \geq \varepsilon > 0,$$

which ensures that the angle between the d^k and the steepest-descent directions $-\nabla f(x^k)$ is uniformly bounded away from $90°$.

In the next section, we explore how each of four very different particular types of minimization methods fits our framework. As part of this exploration, we will see different specialized non-degeneracy conditions ensuring stationarity-inducing target-gaps.

Chapter 2
Particular Reduce-or-Retreat Methods

Four particular examples of methods within our framework are explored in detail, including generalized line-search and trust-region methods, as well as a general pattern-search method and the classic Nelder–Mead method. In each case, we consider descent, and develop conditions ensuring lower-diminishing target-gaps and an approach to stationarity. In the final section, we introduce a degeneracy-check which may be included (as part of an expanded stopping criterion) to ensure that degeneracy is avoided.

The first two examples of methods within our framework that we will explore in detail are very familiar in the context of smooth objective functions $f \in \mathscr{C}^1$. However, our generalized versions below allow non-smooth objectives.

2.1 Generalized Line-Search Method

Line-search methods compute a descent direction d^k at each iteration and carry out a line-search in that direction to obtain a new iterate. In order to qualify as a descent direction, d^k needs to make an angle of no more than $90°$ with some regular subgradient of $-f$ at x^k. Under this requirement, there exists some $v \in \hat{\partial}(-f)(x^k)$ with $\langle v, d^k \rangle \geq 0$, and the direction d^k can always be rescaled to ensure that $v \in \mathbb{B}(0; b\|d^k\|)$ for a fixed $b \in (0, \infty)$. With this in mind, we assume the following:

$$V^k := \left\{ v \in \hat{\partial}(-f)(x^k) \cap \mathbb{B}(0; b\|d^k\|) \text{ with } \langle v, d^k \rangle \geq 0 \right\} \neq \emptyset. \qquad (2.1)$$

Quasi-Newton methods for smooth objective functions $f \in \mathscr{C}^1$ use directions $d^k := (B^k)^{-1} \cdot [-\nabla f(x^k)]$ defined by positive-definite matrices B^k. Since $\hat{\partial}(-f)(x^k) = \{-\nabla f(x^k)\}$ in this case [7, Exercise 8.8], the positive definiteness of the B^k (and the consequent positive definiteness of their inverses) ensures that the set V^k from Eq. (2.1) satisfies $V^k = \{-\nabla f(x^k)\}$, for $b \in (0, \infty)$ any upper bound on the norms of the matrices B^k.

A.B. Levy, *Stationarity and Convergence in Reduce-or-Retreat Minimization*, SpringerBriefs in Optimization, DOI 10.1007/978-1-4614-4642-2_2, © Adam B. Levy 2012

Our generalized line-search method uses a singleton iterate-set $X^k = \{x^k\}$, and a singleton trial-set $T^k = \{x^k + \Delta^k d^k\}$. The reduction test is

$$f(x^k) - f(\xi^k) \geq c\Delta^k \Gamma^k \qquad (2.2)$$

in terms of a fixed value $c \in (0,1)$, and we will use backtracking for retreat. Backtracking is a simple line-search where the trial-size (in this case, step-size) Δ^k is repeatedly reduced by a factor in $[\theta_L, \theta_u]$ until the reduction test (2.2) is satisfied. Since the iterates do not change after retreat, the required condition (1.1) that there is no increase in the f-value of the iterate after retreat is trivially satisfied in this case.

When the objective function is smooth $f \in \mathcal{C}^1$, it is traditional to use the term $\langle -\nabla f(x^k), d^k \rangle$ in place of our target-gap Γ^k in the reduction test (2.2). Our construction allows non-smooth objective functions and general target-gaps Γ^k.

2.1.1 Method Outline

Generalized Line-Search Method

Step 1. Initialize the iterate-set $X^0 = \{x^0\}$ and the trial-size $\Delta^0 = 1$, and choose an initial descent direction d^0 to generate the corresponding trial-set $T^0 = \{x^0 + \Delta^0 d^0\}$. Fix $\theta_L \leq \theta_U$ in $(0,1)$ and $\theta_{\text{red}} = 1$, and set $k = 0$.

Step 2. Stopping criterion.

- (Terminate) If the target-gap Γ^k is zero, then:

  - Set $\bar{x} = x^k$ as the identified target
  - Fix all later iterate-sets, trial-sizes, and target-gaps

$$X^{i+1} = \underbrace{\{\bar{x}, \bar{x}, \dots, \bar{x}\}}_{m \text{ copies}}, \ \Delta^{i+1} = 0, \text{ and } \Gamma^{i+1} = 0 \quad \text{for } i \geq k.$$

- Otherwise, continue to Step 3.

Step 3. Reduce or retreat. Choose $x_{\text{out}}^k = x^k$.

- (Reduce) If the reduction test (2.2) is satisfied for $\xi^k = x^k + \Delta^k d^k$, then:

  - Swap ξ^k for x_{out} in the iterate-set

$$X^{k+1} = \left\{ \xi^k \right\} \cup X^k \setminus \left\{ x_{\text{out}}^k \right\},$$

(continued)

(continued)
- Reset $\Delta^{k+1} = 1$ $(\geq \theta_{\text{red}} \Delta^k)$.
- then continue to Step 4.

- (Retreat) Otherwise, shrink the trial-size $\Delta^{k+1} \in [\theta_L \Delta^k, \theta_U \Delta^k]$, and keep the iterate-set the same $X^{k+1} = X^k$; then continue to Step 4.

Step 4. Compute the new target-gap Γ^{k+1}, and generate the new descent direction d^{k+1} and corresponding new trial-set $T^{k+1} = \{x^{k+1} + \Delta^{k+1} d^{k+1}\}$; increase the iteration-index by 1, and return to Step 2.

2.1.2 Descent

The next result is well known in the special case of a smooth objective $f \in \mathcal{C}^1$ with target-gaps defined by $\Gamma^k = \langle -\nabla f(x^k), d^k \rangle$. Our statement is much more general.

Proposition 5. *For any target-gaps Γ^k, the generalized line-search method is a descent method.*

Proof. After a reduction at the kth iteration, the iterates in this case satisfy $x^{k+1} = \xi^k = x^k + \Delta^k d^k$, so the reduction test (2.2) guarantees that

$$f(x^k) - f\left(x^{k+1}\right) \geq c \Delta^k \Gamma^k.$$

This bound ensures a reduction in the f-value of the iterate $f(x^{k+1}) < f(x^k)$, since c, Δ^k, and Γ^k are all strictly positive (by the assumptions and structure of the method). $\square$

2.1.3 Lower-Diminishing Target-Gaps

The next result is fundamental in the convergence analysis of the generalized line-search method, and it illustrates the significance of the property that the target-gaps are lower-diminishing with trial-sizes.

Proposition 6. *For the generalized line-search method, if the objective function f is bounded below on the set of iterates,*

$$\inf_k f(x^k) > -\infty,$$

and the target-gaps Γ^k are lower-diminishing with the trial-sizes Δ^k, then the target-gaps Γ^k are lower-diminishing.

Proof. If the trial-sizes Δ^k are lower-diminishing, the result follows from the assumption that the Γ^k are lower-diminishing with Δ^k. On the other hand, if the trial-sizes are not lower-diminishing, then they eventually satisfy $\Delta^k \geq \Delta > 0$. Since retreat always shrinks the trial-size by at least the fraction θ_U, we know that there must be a subsequence of reductions in this case. For that subsequence of iterations, we know that the reduction test (2.2) is satisfied

$$f(x^k) - f(x^{k+1}) \geq c\Delta^k \Gamma^k \geq c\Delta \Gamma^k \geq 0.$$

Since the function values never increase at any iteration, and since f is assumed bounded below on the set of iterates, we conclude that there is at least a sub-subsequence with the target-gaps Γ^k converging to zero.   □

According to Proposition 6, the key to deducing lower-diminishing target-gaps is the connection to lower-diminishing trial-sizes. The following corollary shows that this connection can be made when the target-gaps are defined by $\Gamma^k := \langle v^k, d^k \rangle$ for $v^k \in V^k$ from Eq. (2.1).

Corollary 1. *For the generalized line-search method, if the objective function f is bounded below on the set of iterates, and the norms $\|d^k\|$ of the descent directions are uniformly bounded above, then the target-gaps $\Gamma^k := \langle v^k, d^k \rangle$ for $v^k \in V^k$ from Eq. (2.1) are lower-diminishing.*

Proof. The result follows immediately from Propositions 6 and 2 as soon as we show condition (1.4) from Proposition 2. We proceed by contradiction, and assume that $\Gamma^k \geq \gamma > 0$ for all k, with no $\Delta > 0$ triggering $\Delta^{k+1} \geq \Delta^k$ whenever $\Delta^k \leq \Delta$. For the generalized line-search method, the bound $\Delta^{k+1} \geq \Delta^k$ holds if and only if the reduction test (2.2) is satisfied at the kth iteration, in which case, a reduction occurs and the trial-size is reset to $\Delta^{k+1} = 1$. Therefore, without loss of generality, we can assume that there is an uninterrupted sequence of retreats monotonically driving the (positive) trial-sizes Δ^k to zero, fixing the iterates $x^k = \bar{x}$, and (necessarily) violating the reduction test (2.2):

$$f(\bar{x}) - f(\bar{x} + \Delta^k d^k) < c\Delta^k \Gamma^k = c\Delta^k \langle v^k, d^k \rangle, \tag{2.3}$$

for $v^k \in \hat{\partial}(-f)(\bar{x}) \cap \mathbb{B}(0; b\|d^k\|)$. Dividing both sides of this inequality by Δ^k yields

$$\frac{-f(\bar{x} + \Delta^k d^k) - (-f(\bar{x}))}{\Delta^k} < c \langle v^k, d^k \rangle. \tag{2.4}$$

Since the direction norms are assumed to be bounded above (and $\|v^k\| \le b \|d^k\|$), we know that there is an infinite set of iteration-indices $\mathcal{K}$ with $d^k \underset{\mathcal{K}}{\to} \bar{d}$ and $v^k \underset{\mathcal{K}}{\to} \bar{v}$.

Taking the limit (inferior) of inequality (2.4) for this subsequence thus implies the bound

$$d(-f)(\bar{x})(\bar{d}) \le c \left\langle \bar{v}, \bar{d} \right\rangle, \tag{2.5}$$

in terms of the "subderivative" of $-f$ at $\bar{x}$ [7, Definition 8.1]:

$$d(-f)(\bar{x})(\bar{d}) := \liminf_{\tau \to 0^+, d \to \bar{d}} \frac{-f(\bar{x} + \tau d) - (-f(\bar{x}))}{\tau}.$$

According to [7, Exercise 8.4] and the fact that $v^k \in \hat{\partial}(-f)(\bar{x})$, we know that

$$d(-f)(\bar{x})(\bar{d}) \ge \lim_{k \in \mathcal{K}} \left\langle v^k, \bar{d} \right\rangle = \left\langle \bar{v}, \bar{d} \right\rangle.$$

Combining this with the bound (2.5), and recalling that $\left\langle v^k, d^k \right\rangle = \Gamma^k \ge \gamma > 0$, gives

$$0 < \left\langle \bar{v}, \bar{d} \right\rangle \le c \left\langle \bar{v}, \bar{d} \right\rangle,$$

which contradicts $c < 1$.   □

2.1.4 Approaching Stationarity

Just as it is traditional to assume a smooth objective function $f \in \mathscr{C}^1$, and to write the reduction test (2.2) with $\left\langle -\nabla f(x^k), d^k \right\rangle$ in place of our target-gap, it is also traditional to prove that a subsequence of the iterates x^k approaches stationarity $\|\nabla f(x^k)\| \to 0$ under the assumption that the cosine-distances from the steepest-descent directions $-\nabla f(x^k)$ to the descent directions d^k are (uniformly) bounded away from zero:

$$\mathrm{cosdist}\left(-\nabla f(x^k), \{d^k\}\right) := \frac{\left\langle -\nabla f(x^k), d^k \right\rangle}{\| -\nabla f(x^k) \| \, \|d^k\|} \ge \varepsilon > 0.$$

The following is our non-smooth analogue of this traditional result.

Corollary 2. *For the generalized line-search method with $v^k \in V^k$ from Eq. (2.1), if there exists an $\varepsilon > 0$ such that for all k*

$$\mathrm{cosdist}\left(v^k, \{d^k\}\right) := \frac{\left\langle v^k, d^k \right\rangle}{\|v^k\| \, \|d^k\|} \ge \varepsilon, \tag{2.6}$$

then the norms $\|v^k\|$ are lower-diminishing with the target-gaps $\Gamma^k := \left\langle v^k, d^k \right\rangle$.

If, in addition, the objective function f is Lipschitz, regular, and bounded below on an open convex set O containing the iterates x^k, and if the norms $\|d^k\|$ of the descent directions are uniformly bounded above, then the target-gaps $\Gamma^k = \langle v^k, d^k \rangle$ are stationarity-inducing, and a subsequence of the iterates approaches stationarity.

Proof. Since the vectors $v^k \in V^k$ satisfy $\langle v^k, d^k \rangle = \Gamma^k$, we know that whenever the target-gaps Γ^k are lower-diminishing, the cosine-distance bound in Eq. (2.6) implies that the products $\|v^k\| \|d^k\|$ in the denominator must also be lower-diminishing. By the definition (2.1) of the set V^k, we know that $\|v^k\| \leq b \|d^k\|$ which (multiplying both sides of this inequality by $\|v^k\|$) then implies that the norms $\|v^k\|$ are also lower-diminishing.

The second part of the result follows from Corollary 1, together with the fact that the regular subgradients $v^k \in V^k$ are automatically subgradients [7, Theorem 8.6]:

$$v^k \in \hat{\partial}(-f)(x^k) \subseteq \partial(-f)(x^k),$$

as well as the resulting inclusion $-v^k \in -\partial(-f)(x^k) \subseteq \partial f(x^k)$ guaranteed by [7, Corollary 9.21] under the additional assumptions on f. $\square$

Remark 6. Notice that under the sufficient cosine-distance condition

$$\mathrm{cosdist}\left(d^k, \hat{\partial}(-f)(x^k) \cap \mathbb{B}\big(0; b\|d^k\|\big)\right) \geq \varepsilon > 0, \tag{2.7}$$

there always exists a $v^k \in V^k$ from Eq. (2.1) satisfying the cosine-distance bound in Eq. (2.6). For smooth objective functions $f \in \mathscr{C}^1$, this non-degeneracy condition (2.7) on the trial-sets reduces to the condition that the angle between the descent directions d^k and the steepest-descent directions $-\nabla f(x^k)$ is uniformly bounded away from $90°$ (as long as the d^k are scaled appropriately).

Remark 7. The Lipschitz and regularity assumptions on the objective function are only necessary to ensure the containment $-\partial(-f)(x^k) \subseteq \partial f(x^k)$ via [7, Corollary 9.21]. If this containment happens to hold anyway, the Lipschitz and regularity assumptions can be dropped.

2.2 Generalized Trust-Region Method

Trust-region methods use a set $X^k \subseteq \mathbb{R}^n$ of points (the iterate-set) to construct a *model function* $\mathrm{mod}^k(p)$ representing the objective function; and then "approximately minimize" $\mathrm{mod}^k(p)$ over a ball with radius equal to the trial-size Δ^k (a trust radius) to generate trial points. We will investigate a generalized trust-region method where the trial-sets have the form

$$T^k = \left\{ \xi \in \mathbb{B}(x^k; \Delta^k) : \mathrm{modred}^k(\xi) \geq \Delta^k \Gamma^k \right\} \tag{2.8}$$

in terms of the target-gaps Γ^k and where

$$\text{modred}^k(\xi) := \text{mod}^k(0) - \text{mod}^k(\xi - x^k)$$

measures the "model-reduction" achieved by moving from $p = 0$ to $p = \xi - x^k$.

Model functions are $\mathscr{C}^2$ and satisfy $\text{mod}^k(0) = f(x^k)$; a classic example for smooth objectives $f \in \mathscr{C}^1$ is a "quasi-Taylor" quadratic of the form

$$\text{mod}^k(p) := f(x^k) + \langle \nabla f(x^k), p \rangle + \frac{1}{2} \langle p, \nabla^2 \text{mod}^k(0) \cdot p \rangle. \tag{2.9}$$

Another useful model function is an interpolating polynomial centered at the iterate x^k:

$$\text{mod}^k(x - x^k) = f(x) \text{ for all } x \in X^k.$$

Because the model gradients and Hessians at zero play a prominent role in both of these cases (and in others), we use the convenient notations

$$g^k := \nabla \text{mod}^k(0) \quad \text{and} \quad H^k := \nabla^2 \text{mod}^k(0)$$

(g for "gradient" and H for "Hessian").

The reduction test for the generalized trust-region method is

$$f(x^k) - f(\xi^k) \geq c_U \Delta^k \Gamma^k \tag{2.10}$$

in terms of a fixed $c_U \in (0,1)$; and the outgoing point $x_{\text{out}}^k \in X^k$ has maximal distance from the iterate x^k. Retreat involves three possible updating schemes, depending on how badly the reduction test is violated (using $c_L \in [0, c_U)$), as well as on the distance between the trial point ξ^k and the iterate x^k. As we will show in Lemma 3 below, none of the three retreat schemes increases the f-value of the iterate; so the required condition (1.1) is satisfied.

For the trial-set T^k (2.8) to be nonempty, the target-gaps Γ^k must satisfy

$$\Gamma^k \leq \max_{\xi \in \mathbb{B}(x^k; \Delta^k)} \frac{\text{modred}^k(\xi)}{\Delta^k}. \tag{2.11}$$

One possible target-gap Γ^k is the maximum in Eq. (2.11), in which case the trial-set consists of the iterate x^k added to the minimizers of the model function $\text{mod}^k(p)$ over $\mathbb{B}(0; \Delta^k)$:

$$T^k = \operatorname*{argmax}_{\xi \in \mathbb{B}(x^k; \Delta^k)} \text{modred}^k(\xi) = \{x^k\} + \operatorname*{argmin}_{p \in \mathbb{B}(0; \Delta^k)} \text{mod}^k(p).$$

When the model function is quadratic

$$\mathrm{mod}^k(p) := f(x^k) + \left\langle g^k, p \right\rangle + \frac{1}{2} \left\langle p, H^k \cdot p \right\rangle,$$

another possible choice of target-gap is

$$\Gamma^k := \frac{1}{2} \|g^k\| \min\left\{ \frac{\|g^k\|}{\Delta^k \|H^k\|}, 1 \right\}, \tag{2.12}$$

which we confirm in the following lemma.

Lemma 2. *When the model function is quadratic, the target-gaps Γ^k defined by Eq. (2.12) satisfy the condition (2.11) ensuring a non-empty trial-set T^k.*

Proof. We prove this by evaluating the model-reduction function in this case

$$\mathrm{modred}^k(\xi) = -\left\langle g^k, \xi - x^k \right\rangle - \frac{1}{2} \left\langle \xi - x^k, H^k \cdot \xi - x^k \right\rangle$$

at the point

$$\bar{\xi}^k := x^k - \Delta^k \frac{g^k}{\|g^k\|} \min\left\{ \frac{\|g^k\|^3}{\Delta^k \langle g^k, H^k \cdot g^k \rangle}, 1 \right\},$$

which by definition satisfies $\bar{\xi}^k \in \mathbb{B}\left(x^k; \Delta^k\right)$. We consider the two cases identified by the min in the definition of $\bar{\xi}^k$.

In the case that $\|g^k\|^3 \le \Delta^k \langle g^k, H^k \cdot g^k \rangle$, we have

$$\begin{aligned}
\mathrm{modred}^k(\bar{\xi}^k) &= \frac{\|g^k\|^4}{\langle g^k, H^k \cdot g^k \rangle} - \frac{1}{2} \left\langle \frac{g^k \|g^k\|^2}{\langle g^k, H^k \cdot g^k \rangle}, H^k \cdot \frac{g^k \|g^k\|^2}{\langle g^k, H^k \cdot g^k \rangle} \right\rangle \\
&= \frac{\|g^k\|^4}{\langle g^k, H^k \cdot g^k \rangle} - \frac{1}{2} \frac{\|g^k\|^4}{\langle g^k, H^k \cdot g^k \rangle} \\
&= \frac{1}{2} \frac{\|g^k\|^4}{\langle g^k, H^k \cdot g^k \rangle} \\
&\ge \frac{1}{2} \frac{\|g^k\|^4}{\|g^k\|^2 \|H^k\|} = \frac{1}{2} \frac{\|g^k\|^2}{\|H^k\|} \\
&\ge \Delta^k \frac{1}{2} \|g^k\| \min\left\{ \frac{\|g^k\|}{\Delta^k \|H^k\|}, 1 \right\} = \Delta^k \Gamma^k.
\end{aligned}$$

From this and the fact that $\bar{\xi}^k \in \mathbb{B}\left(x^k; \Delta^k\right)$, it follows that Γ^k satisfies Eq. (2.11) in this case.

If instead $\|g^k\|^3 > \Delta^k \langle g^k, H^k \cdot g^k \rangle$, we have

$$
\begin{aligned}
\mathrm{modred}^k(\bar{\xi}^k) &= \Delta^k \|g^k\| - \frac{1}{2} \left\langle \Delta^k \frac{g^k}{\|g^k\|}, H^k \cdot \Delta^k \frac{g^k}{\|g^k\|} \right\rangle \\
&= \Delta^k \|g^k\| - \frac{1}{2} \frac{\Delta^k}{\|g^k\|^2} \Delta^k \langle g^k, H^k \cdot g^k \rangle \\
&\geq \Delta^k \|g^k\| - \frac{1}{2} \frac{\Delta^k}{\|g^k\|^2} \|g^k\|^3 = \Delta^k \frac{1}{2} \|g^k\| \\
&\geq \Delta^k \frac{1}{2} \|g^k\| \min \left\{ \frac{\|g^k\|}{\Delta^k \|H^k\|}, 1 \right\} = \Delta^k \Gamma^k.
\end{aligned}
$$

As above, we conclude that Γ^k satisfies Eq. (2.11) in this case also. $\qquad\square$

2.2.1 Method Outline

Generalized Trust-Region Method

Step 1. Initialize the iterate-set X^0 and the trial-size $\Delta^0 > 0$, and generate the corresponding trial-set T^0 from (2.8). Fix $\theta_L \leq \theta_U$ in $(0,1)$ and $\theta_{\mathrm{red}} \geq 1$, and set $k = 0$

Step 2. Stopping criterion.

- (Terminate) If the target-gap Γ^k is zero, then:

  - Set $\bar{x} = x^k$ as the identified target.
  - Fix all later iterate-sets, trial-sizes, and target-gaps

  $$
  X^{i+1} = \underbrace{\{\bar{x}, \bar{x}, \ldots, \bar{x}\}}_{m \text{ copies}}, \Delta^{i+1} = 0, \text{ and } \Gamma^{i+1} = 0 \quad \text{for } i \geq k.
  $$

- Otherwise, continue to Step 3.

Step 3. Reduce or retreat. Choose $x_{\mathrm{out}}^k \in \operatorname*{argmax}_{x \in X^k} \|x^k - x\|$ and $\xi^k \in T^k$.

- (Reduce) If the reduction test (2.10) is satisfied, then

  - Swap ξ^k for x_{out} in the iterate-set

  $$
  X^{k+1} = \left\{ \xi^k \right\} \cup X^k \setminus \left\{ x_{\mathrm{out}}^k \right\}.
  $$

(continued)

(continued)
- Choose $\Delta^{k+1} \geq \theta_{\text{red}} \Delta^k$.
- Then continue to Step 4.

- (Retreat) Otherwise, shrink the trial-size $\Delta^{k+1} \in [\theta_L \Delta^k, \theta_U \Delta^k]$, and:

  - If either (i) there is an almost-sufficient reduction,

$$f(x^k) - f(\xi^k) \geq c_L \Delta^k \Gamma^k, \tag{2.13}$$

  or (ii) there is increase $f(x^k) - f(\xi^k) < 0$ and $\|\xi^k - x^k\| \leq \|x_{\text{out}}^k - x^k\|$, then swap ξ^k for x_{out} in the iterate-set

$$X^{k+1} = \left\{ \xi^k \right\} \cup X^k \setminus \left\{ x_{\text{out}}^k \right\};$$

  then continue to Step 4.
  - For any other case, keep the same iterate-set $X^{k+1} = X^k$; then continue to Step 4.

Step 4. Construct the new model function $\text{mod}^{k+1}(p)$, compute the new target-gap Γ^{k+1}, and generate the new trial-set T^{k+1} from (2.8); increase the iteration-index by 1, and return to Step 2.

The first retreat option allows the iterate-set to change "for the better" in some sense, by swapping in the trial point ξ^k if it provides an almost-sufficient reduction, or if it retracts the iterate-set. In any case, as the next lemma shows, retreat does not increase the f-value of the iterate (which is required by our framework).

Lemma 3. *The generalized trust-region method never increases the f-value of the iterates after retreat.*

Proof. If the kth iteration induces the "any other case" retreat alternative, we have $x^k = x^{k+1}$ so there is trivially no increase in the f-value of the iterate. Otherwise, if retreat is induced at the kth iteration, there are two cases to consider.

In the first case (i), almost-sufficient reduction (2.13) guarantees that the f-value of the iterate is reduced $f(\xi^k) < f(x^k)$ at the trial point ξ^k, since in that case, c_L, Δ^k, and Γ^k are strictly positive. Since the trial point ξ^k is added to the iterate-set in this case, we have

$$f(x^{k+1}) = \min_{x \in X^{k+1}} f(x) \leq f(\xi^k) < f(x^k).$$

In the second case (ii), where $f(x^k) - f(\xi^k) < 0$, we know in particular that $\xi^k \neq x^k$. Thus, the inequality $\|\xi^k - x^k\| \leq \|x^k_{\text{out}} - x^k\|$ implies that the outgoing point x^k_{out} is not the iterate x^k either. Hence, $x^k \in X^{k+1}$ and

$$f(x^{k+1}) = \min_{x \in X^{k+1}} f(x) \leq f(x^k). \qquad \square$$

Remark 8. Notice that retreat in case (ii) does not necessarily produce a trial point ξ^k with lower f-value than the point x^k_{out} it replaces in the iterate-set. Recall that this property holds for any iteration that induces *reduction*; but evidently does not always hold for retreat.

2.2.2 Descent

The following result is a little more surprising than its analogue in the case of the generalized line-search method since there are no explicit descent directions involved here.

Proposition 7. *For any target-gaps Γ^k satisfying Eq. (2.11), the generalized trust-region method is a descent method.*

Proof. If the kth iteration induces a reduction, the reduction test (2.10) guarantees that the f-value of the iterate is reduced $f(\xi^k) < f(x^k)$ at the trial point ξ^k, since in that case, c_U, Δ^k, and Γ^k are strictly positive. Since the trial point ξ^k is added to the iterate-set in this case, we have

$$f(x^{k+1}) = \min_{x \in X^{k+1}} f(x) \leq f(\xi^k) < f(x^k). \qquad \square$$

2.2.3 Lower-Diminishing Target-Gaps

The next result is fundamental in the convergence analysis of the generalized trust-region method, and it has a similar proof to that of its analogue Proposition 6 for the generalized line-search method.

Proposition 8. *For the generalized trust-region method, if the objective function f is bounded below on the set of iterates,*

$$\inf_k f(x^k) > -\infty,$$

and the target-gaps Γ^k are lower-diminishing with the trial-sizes Δ^k, then the target-gaps Γ^k are lower-diminishing.

Proof. If the trial-sizes Δ^k are lower-diminishing, the result follows from the assumption that the Γ^k are lower-diminishing with Δ^k. On the other hand, if the trial-sizes are not lower-diminishing, then $\Delta^k \geq \Delta > 0$ for all k eventually. Since retreat always shrinks the trial-size by at least the fraction θ_U, we know that there must be a subsequence of reductions in this case. For that subsequence of iterations, we know that the reduction test (2.10) is satisfied

$$f(x^k) - f(\xi^k) \geq c_U \Delta^k \Gamma^k \geq c_u \Delta \Gamma^k > 0,$$

and that the next iterate $x^{k+1} \in \operatorname*{argmin}_{x \in X^{k+1}} f(x)$ satisfies $f(x^{k+1}) \leq f(\xi^k)$, since $\xi^k \in X^{k+1}$. Putting these two facts together gives

$$f(x^k) - f(x^{k+1}) \geq c_U \Delta^k \Gamma^k \geq c_u \Delta \Gamma^k > 0.$$

Since the function values never increase at any iteration, and since f is assumed bounded below on the set of iterates, we conclude that there is at least a sub-subsequence with the target-gaps Γ^k converging to zero.   □

As we saw with the generalized line-search method, the key is to establish that the target-gaps Γ^k are lower-diminishing with the trial-sizes Δ^k. The next corollary gives conditions which ensure this, and uses scalars $G \in (0, \infty)$ and $H \in (0, \infty)$ in bounds on the model gradients g^k (hence "G") and Hessians H^k (hence "H"), respectively.

Corollary 3. *For the generalized trust-region method and any target-gaps Γ^k satisfying Eq. (2.11), suppose that for every $\gamma > 0$, there exist $\Delta > 0$, $G \in (0, \infty)$, and $H \in (0, \infty)$ satisfying*

$$\Delta (G + H) < \gamma (1 - c_U), \tag{2.14}$$

and such that whenever $\Gamma^k \geq \gamma$ and $\Delta^k \leq \Delta$, we have that

- *The model gradient $g^k := \nabla \mathrm{mod}^k(0)$ satisfies*

$$\sup \left\{ \|v - g^k\| \,\middle|\, v \in \bigcup_{x \in \mathbb{B}(x^k; \Delta^k)} \partial f(x) \cup -\partial(-f)(x) \right\} \leq G\Delta \tag{2.15}$$

- *f is bounded-below on the set of iterates, and Lipschitz continuous on an open convex set O containing $\bigcup_k \mathbb{B}(x^k; \Delta)$.*
- *The model Hessian $\nabla^2 \mathrm{mod}^k(p)$ satisfies*

$$\sup_{p \in \mathbb{B}(0; \Delta^k)} \|\nabla^2 \mathrm{mod}^k(p)\| \leq H. \tag{2.16}$$

Then the target-gaps Γ^k are lower-diminishing.

Proof. The result follows immediately from Propositions 8 and 2 as soon as we show condition (1.4) from Proposition 2. Accordingly, we assume that $\Gamma^k \geq \gamma > 0$ for all k, and demonstrate that $\Delta^k \leq \Delta$ implies $\Delta^{k+1} \geq \Delta^k$. Since $\Gamma^k \geq \gamma > 0$, we know that termination is avoided, and so the trial-sizes only satisfy $\Delta^{k+1} \geq \Delta^k$ after reduction. Consequently, we need $\Delta^k \leq \Delta$ to imply that the reduction test (2.10) is satisfied at the kth iteration. Based on the construction of the trial-set T^k (2.8), the reduction test (2.10) is implied by the inequality

$$f(x^k) - f(\xi^k) \geq c_U \, \mathrm{modred}^k(\xi^k). \tag{2.17}$$

The same construction implies the bounds that

$$\mathrm{modred}^k(\xi^k) \geq \Delta^k \Gamma^k \geq \Delta^k \gamma > 0. \tag{2.18}$$

Dividing both sides of Eq. (2.17) by the model-reduction term then gives the equivalent expression

$$\frac{f(x^k) - f(\xi^k)}{\mathrm{modred}^k(\xi^k)} \geq c_U,$$

which is itself implied by the bound

$$\left| \frac{f(x^k) - f(\xi^k)}{\mathrm{modred}^k(\xi^k)} - 1 \right| \leq 1 - c_U. \tag{2.19}$$

Thus we will be done if we can show that $\Delta^k \leq \Delta$ implies the bound (2.19).

Note that the term on the left side of Eq. (2.19) satisfies

$$\left| \frac{f(x^k) - f(\xi^k)}{\mathrm{modred}^k(\xi^k)} - 1 \right| = \left| \frac{\mathrm{mod}^k(\xi^k - x^k) - f(\xi^k)}{\mathrm{modred}^k(\xi^k)} \right|$$

$$\leq \frac{\left| \mathrm{mod}^k(\xi^k - x^k) - f(\xi^k) \right|}{\Delta^k \gamma} \tag{2.20}$$

since $\mathrm{mod}^k(0) = f(x^k)$ and $\mathrm{modred}^k(\xi^k) \geq \Delta^k \gamma$ from Eq. (2.18). To make headway from here, we construct a quadratic Taylor expansion of the model function mod^k about 0:

$$\mathrm{mod}^k(p) = f(x^k) + \left\langle g^k, p \right\rangle + \left\langle p, \widetilde{H}^k \cdot p \right\rangle$$

where $\widetilde{H}^k$ is the Hessian of mod^k evaluated at some intermediate point in $\mathbb{B}(0; \|p\|)$. Substituting $p = \xi^k - x^k$ and subtracting $f(\xi^k)$ from this Taylor expansion give

$$\left| \mathrm{mod}^k(\xi^k - x^k) - f(\xi^k) \right| \leq \left| f(x^k) - f(\xi^k) + \left\langle g^k, \xi^k - x^k \right\rangle \right|$$

$$+ \|\widetilde{H}^k\| \left(\Delta^k \right)^2, \tag{2.21}$$

since $\xi^k \in \mathbb{B}(x^k; \Delta^k)$.

The assumed Lipschitz continuity of f on O implies via [7, Theorem 10.48] that there exist $\tilde{v}^k \in \partial f(\tilde{x}^k) \cup -\partial(-f)(\tilde{x}^k)$ for $\tilde{x}^k \in \mathbb{B}(x^k; \Delta^k)$ with

$$f(x^k) - f(\xi^k) = \left\langle \tilde{v}^k, x^k - \xi^k \right\rangle. \tag{2.22}$$

We can substitute this into Eq. (2.21) to get

$$\left| \text{mod}^k(\xi^k - x^k) - f(\xi^k) \right| \leq \|\tilde{v}^k - g^k\| \Delta^k + \|\tilde{H}^k\| (\Delta^k)^2.$$

If we divide both sides by $\Delta^k \gamma$ and combine with Eq. (2.20) we get the inequality

$$\left| \frac{f(x^k) - f(\xi^k)}{\text{modred}^k(\xi^k)} - 1 \right| \leq \frac{\|\tilde{v}^k - g^k\|}{\gamma} + \frac{\|\tilde{H}^k\| \Delta^k}{\gamma}. \tag{2.23}$$

The bound (2.15) implies that the first term on the right side of Eq. (2.23) is bounded above by $\frac{G\Delta}{\gamma}$, and the bound (2.16) on the model Hessians implies that the second term on the right side of Eq. (2.23) is bounded above by $\frac{H\Delta^k}{\gamma} \leq \frac{H\Delta}{\gamma}$:

$$\left| \frac{f(x^k) - f(\xi^k)}{\text{modred}^k(\xi^k)} - 1 \right| \leq \frac{G\Delta}{\gamma} + \frac{H\Delta}{\gamma}.$$

This, together with Eq. (2.14), gives us the bound (2.19) from which our result follows.   □

Remark 9. The bound (2.15) is more general than, but related to the "Taylor-like" bounds obtained for interpolation polynomials and smooth objective functions in [2].

Remark 10. A similar argument shows that the product $\Delta^k \Gamma^k$ is lower-diminishing under the weaker assumptions that

$$\|g^k\| \leq G \quad \text{and} \quad \sup_{p \in \mathbb{B}(0; \Delta^k)} \|\nabla^2 \text{mod}^k(p)\| \Delta^k \leq 1$$

replacing Eqs. (2.15) and (2.16) in Corollary 3. The first of these bounds is satisfied by the classic model functions (2.9) where $g^k = \nabla f(x^k)$ under the assumption of bounded iterates; and the second bound is weaker than the bound (2.16) and in particular allows unbounded Hessians as long as the trial-sizes Δ^k approach zero quickly enough.

The only significant change in the argument involves the bounds starting from Eq. (2.23) and uses [7, Theorem 9.13] and [7, Proposition 5.15] applied to the mapping

$$S(x) := \begin{cases} \partial f(x) \cup -\partial(-f)(x) & \text{if } x \in O \\ \{0\} & \text{otherwise} \end{cases}$$

to deduce that all of the $\|\tilde{v}^k\|$ are bounded above by some constant $V \in (0, \infty)$.

Remark 11. When f is regular, the bound (2.15) simplifies to

$$\sup\left\{\|v-g^k\| \,\middle|\, v \in \bigcup_{x\in\mathbb{B}\left(x^k;\Delta^k\right)} \partial f(x)\right\} \le G\Delta \tag{2.24}$$

(via [7, Theorem 10.48]). In the next section, we show that a slightly stronger bound (with Δ^k replacing Δ on the right) helps ensure stationarity-inducing target-gaps.

2.2.4 Approaching Stationarity

If we strengthen the bound (2.24) by replacing Δ with $\Delta^k (\le \Delta)$, we can deduce stationarity-inducing target-gaps.

Proposition 9. *For the generalized trust-region method and any target-gaps Γ^k satisfying Eq. (2.11), suppose that for every $\gamma > 0$, there exist $\Delta > 0$ and $G \in (0, \infty)$ such that whenever $\Delta^k \le \Delta$ and $\Gamma^k \ge \gamma$, the model gradient $g^k := \nabla\mathrm{mod}^k(0)$ satisfies*

$$\sup\left\{\|v-g^k\| \,\middle|\, v \in \bigcup_{x\in\mathbb{B}\left(x^k;\Delta^k\right)} \partial f(x)\right\} \le G\Delta^k, \tag{2.25}$$

and f is Lipschitz continuous on an open convex set O containing $\bigcup_k \mathbb{B}(x^k; \Delta)$. If the trial-sizes Δ^k converge to zero, and the model gradient-norms $\|g^k\|$ are lower-diminishing with the target-gaps Γ^k, then the target-gaps are stationarity-inducing.

If in addition, the objective f is regular and bounded below on the set of iterates, and there exists $H \in (0, \infty)$ satisfying Eq. (2.16), then a subsequence of the iterates x^k approaches stationarity.

Proof. Since f is Lipschitz on O containing all the iterates x^k, we know in particular that $\partial f(x^k)$ is non-empty for all k [7, Theorem 9.61]. If we choose any $v^k \in \partial f(x^k)$, we can apply the bound (2.25) to deduce that $\|v^k - g^k\| \le G\Delta^k$ whenever $\Delta^k \le \Delta$. Thus, since the trial-sizes converge to zero, we know that eventually the subgradients v^k are arbitrarily close to the model gradients g^k.

When the target-gaps Γ^k are lower-diminishing, we know that the model gradient-norms $\|g^k\|$ are also lower-diminishing (by assumption). It follows that the subgradient-norms $\|v^k\|$ are also lower-diminishing, so that the target-gaps are stationarity-inducing.

Under the additional assumptions f and $H \in (0, \infty)$, we can apply Corollary 3 (via the simplified bound (2.24) in this case) to deduce that the target-gaps Γ^k are lower-diminishing (Note that we can shrink $\Delta > 0$ if necessary to ensure that Eq. (2.14) is satisfied.). Then the preceding argument ensures that the subgradient-norms $\|v^k\|$ are lower-diminishing too. $\qquad\square$

Remark 12. Under the assumptions (2.16) and $\Delta^k \leq \Delta$, the target-gaps (2.12) satisfy the property that the model gradient-norms are lower-diminishing with Γ^k.

Remark 13. The bound (2.25) is a non-degeneracy condition on the trial-sets T^k (2.8) associated with the generalized trust-region method, since the trial-sets T^k (2.8) are determined by the model function $\text{mod}^k(p)$ (via the model reduction function). Thus, the bound (2.25) ensures that the $\text{mod}^k(p)$ appropriately models the objective function f by ensuring that the model gradients g^k are close enough to the objective subgradients on $\mathbb{B}(x^k; \Delta^k)$. For a smooth objective function $f \in \mathscr{C}^1$, the bound (2.25) entails the model gradients g^k being within $G\Delta^k$ (in norm) of all the objective gradients $\nabla f(x)$ for $x \in \mathbb{B}(x^k; \Delta^k)$.

So far, we've described two examples of generalizations of very well-known particular reduce-or-retreat methods. The next method we'll describe is a generalization of a pattern-search method from [8], where we allow very general patterns and also include an explicit stopping criterion in terms of a general target-gap Γ^k.

2.3 General Pattern-Search Method

As in the case of our generalized line-search method, the iterate-sets in our general pattern-search method are singletons $X^k = \{x^k\}$, and the trial-sizes Δ^k are step-sizes. The trial-sets T^k are defined by

$$T^k = \left\{ x^k + \Delta^k d^k \,\middle|\, d^k \in \text{col}\left(D^k\right) \right\} \tag{2.26}$$

in terms of the set $\text{col}\left(D^k\right)$ of nonzero column vectors associated with some "pattern-matrix" $D^k \in \mathbb{Z}^{n \times d}$.

The reduction test in this case is $f(\xi^k) < f(x^k)$, and retreat shrinks the trial-size with no changes to the iterate-set. Since the iterates do not change after retreat, the required condition (1.1) that there is no increase in the f-value of the iterate after retreat is trivially satisfied in this case.

2.3.1 Method Outline

General Pattern-Search Method

Step 1. Initialize the iterate-set $X^0 = \{x^0\}$ and the trial-size $\Delta^0 = 1$, and choose the initial pattern-matrix D^0 to generate the corresponding trial-set T^0 from (2.26). Fix $\theta_L \leq \theta_U$ in $(0, 1)$ and $\theta_{\text{red}} \geq 1$, and set $k = 0$.

(continued)

(continued)

Step 2. Stopping criterion.

- (Terminate) If the target-gap Γ^k is zero, then:

  - Set $\bar{x} = x^k$ as the identified target.
  - Fix all later iterate-sets, trial-sizes, and target-gaps

  $$X^{i+1} = \underbrace{\{\bar{x}, \bar{x}, \ldots, \bar{x}\}}_{m \text{ copies}}, \Delta^{i+1} = 0, \text{ and } \Gamma^{i+1} = 0 \quad \text{for } i \geq k.$$

- Otherwise, continue to Step 3.

Step 3. Reduce or retreat. Choose $x_{\text{out}}^k = x^k$.

- (Reduce) If there exists a trial point $\xi^k \in T^k$ with $f(\xi^k) < f(x^k)$, then,

  - swap ξ^k for x_{out} in the iterate-set

  $$X^{k+1} = \left\{\xi^k\right\} \cup X^k \setminus \left\{x_{\text{out}}^k\right\},$$

  - Choose $\Delta^{k+1} \geq \theta_{\text{red}} \Delta^k$,
  - Then continue to Step 4.

- (Retreat) Otherwise, shrink the trial-size $\Delta^{k+1} \in [\theta_L \Delta^k, \theta_U \Delta^k]$, and keep the iterate-set the same $X^{k+1} = X^k$; then continue to Step 4.

Step 4. Compute the new target-gap Γ^{k+1}, and choose the new pattern-matrix D^{k+1} to generate the new trial-set T^{k+1} from (2.26); increase the iteration-index by 1, and return to Step 2.

2.3.2 Descent

The next result follows immediately from the construction of the method.

Proposition 10. *For any target-gaps Γ^k, the general pattern-search method is a descent method.*

Proof. By construction, the iterates in this case satisfy $f(x^{k+1}) = f(\xi^k) < f(x^k)$ after a reduction at the kth iteration. □

2.3.3 Lower-Diminishing Target-Gaps

We can deduce lower-diminishing target-gaps Γ^k from our Proposition 2.

Proposition 11. *For the general pattern-search method, any target-gaps Γ^k satisfying*

$$\Gamma^k \geq \gamma > 0 \,\forall k \;\Rightarrow\; \exists \Delta > 0 \text{ with } f(x^k) > \min_{\xi \in T^k} f(\xi) \text{ whenever } \Delta^k \leq \Delta, \quad (2.27)$$

are lower-diminishing with the trial-sizes Δ^k.

Proof. This follows immediately from Proposition 2, since the bound $f(x^k) > \min_{\xi \in T^k} f(\xi)$ in the condition (2.27) guarantees a reduction (and hence a nondecreasing trial-size) whenever $\Delta^k \leq \Delta$.   □

Remark 14. Notice that when $\Gamma^k \geq \gamma > 0$, the bound

$$f(x^k) \geq \min_{\xi \in T^k} f(\xi) + c\,\Gamma^k \Delta^k, \quad (2.28)$$

in terms of a fixed $c \in (0, \infty)$, implies the bound $f(x^k) > \min_{\xi \in T^k} f(\xi)$ in condition (2.27). If the initial level set $\mathrm{lev}_{f(x^0)}\, f$ is compact, the bound (2.28) is satisfied by target-gaps $\Gamma^k := \|\nabla f(x^k)\|$ when $f \in \mathscr{C}^1$ in the pattern-search method of [8] (via [8, Eqs. (15) and (23)]). Since [8, Theorem 3.3] shows that the trial-sizes are lower-diminishing in that case, one implication of Proposition 11 is a major convergence result in [8] that a subsequence of the iterates x^k approaches stationarity when $f \in \mathscr{C}^1$.

2.3.4 Approaching Stationarity

The following corollary shows that the iterates of our general pattern-search method approach stationarity whenever the trial-sizes are lower-diminishing, provided that the sufficient cosine-distance condition holds

$$\inf_{v \in \partial f(x^k)} \mathrm{cosdist}\left(v, \mathrm{col}\left(D^k\right)\right) := \inf_{v \in \partial f(x^k)} \max_{d^k \in \mathrm{col}(D^k)} \frac{\langle v, d^k \rangle}{\|v\|\,\|d^k\|} \geq \varepsilon, \quad (2.29)$$

for some $\varepsilon > 0$ independent of k, together with the bound

$$\inf_{v \in \partial f(x^k)} \left\{ \sup \|\tilde{v} - v\| \;\middle|\; \tilde{v} \in \bigcup_{x \in \mathbb{B}\left(x^k; \Delta^k\right)} \partial f(x) \cup -\partial(-f)(x) \right\} \leq G\Delta \quad (2.30)$$

on the subgradients. When $f \in \mathscr{C}^1$, the bound (2.30) reduces to

$$\|\nabla f(x) - \nabla f(x^k)\| \leq G\Delta \quad \text{for} \quad x \in \mathbb{B}(x^k; \Delta^k)$$

in terms the gradient near x^k.

Corollary 4. *For the general pattern-search method with the target-gaps $\Gamma^k :=$* $\inf\limits_{v \in \partial f(x^k)} \|v\|$, *suppose that for every $\gamma > 0$, there exist $\Delta > 0$, $\varepsilon > 0$, and $G \in (0, \infty)$* *satisfying*

$$G\Delta < \varepsilon\gamma, \tag{2.31}$$

and such that whenever $\Delta^k \leq \Delta$ and $\Gamma^k \geq \gamma$, f is Lipschitz continuous on an open *convex set O containing $\bigcup_k \mathbb{B}(x^k; \Delta^k)$, and conditions (2.29) and (2.30) hold. Then* *a subsequence of the iterates x^k approaches stationarity if the trial-sizes are lower-* *diminishing.*

Proof. These target-gaps Γ^k are trivially stationarity-inducing, so the result follows from Proposition 11, as soon as we show that the condition (2.27) is satisfied. Accordingly, we assume $\Gamma^k \geq \gamma$ for all k.

Since f is Lipschitz on O containing all the iterates x^k, we know in particular that the set of subgradients $\partial f(x^k)$ is non-empty [7, Theorem 9.61], closed [7, Theorem 8.6], and bounded [7, Theorem 9.13]. Hence, we can choose $v^k \in \partial f(x^k)$ satisfying the bound (2.30) on the subgradients:

$$\left\{ \sup \|\tilde{v} - v^k\| \,\middle|\, \tilde{v} \in \bigcup_{x \in \mathbb{B}(x^k; \Delta^k)} \partial f(x) \cup -\partial(-f)(x) \right\} \leq G\Delta. \tag{2.32}$$

We now apply the sufficient cosine-distance condition (2.29) to choose a trial point ξ^k from the trial-set T^k (2.26) satisfying

$$\frac{\langle v^k, x^k - \xi^k \rangle}{\|v^k\| \, \|x^k - \xi^k\|} \geq \varepsilon. \tag{2.33}$$

The Lipschitz continuity of f on O implies via [7, Theorem 10.48] that there exist $\tilde{v}^k \in \partial f(\tilde{x}^k) \cup -\partial(-f)(\tilde{x}^k)$ for $\tilde{x}^k \in \mathbb{B}(x^k; \Delta^k)$ with

$$f(x^k) - f(\xi^k) = \langle \tilde{v}^k, x^k - \xi^k \rangle.$$

By adding and subtracting the term $\langle v^k, x^k - \xi^k \rangle$ from the right side of this equation, and then applying the bound (2.33), we get

$$f(x^k) - f(\xi^k) = \left\langle \tilde{v}^k - v^k, x^k - \xi^k \right\rangle + \left\langle v^k, x^k - \xi^k \right\rangle$$

$$\geq \left\langle \tilde{v}^k - v^k, x^k - \xi^k \right\rangle + \varepsilon \, \|v^k\| \, \|x^k - \xi^k\|$$

$$\geq \left(\varepsilon \gamma - \|\tilde{v}^k - v^k\| \right) \|x^k - \xi^k\|$$

$$\geq (\varepsilon \gamma - G\Delta) \, \|x^k - \xi^k\|,$$

where we have used $\Gamma^k = \inf_{v \in \partial f(x^k)} \|v\| \geq \gamma > 0$ in the second-to-last inequality and we have applied Eq. (2.32) in the final inequality. We conclude from this and the bound (2.31) that $f(x^k) > f(\xi^k)$, so that the condition (2.27) is satisfied.   □

Remark 15. When f is regular, the bound (2.30) on the subgradients simplifies to

$$\inf_{v \in \partial f(x^k)} \left\{ \sup \|\tilde{v} - v\| \;\middle|\; \tilde{v} \in \bigcup_{x \in \mathbb{B}\left(x^k; \Delta^k\right)} \partial f(x) \right\} \leq G\Delta.$$

Remark 16. The pattern-matrices in [8] are structured so that their columns "positively span" $\mathbb{R}^n$ (i.e., non-negative linear combinations of the columns span $\mathbb{R}^n$) in such a way that sufficient cosine-distances can be made from any vector $v \in \mathbb{R}^n$ (via [8, Lemma 6.2]):

$$\inf_{v \in \mathbb{R}^n} \operatorname{cosdist}\left(v, \operatorname{col}\left(D^k \right) \right) \geq \varepsilon.$$

Under the assumption of a compact initial level set $\operatorname{lev}_{f(x^0)} f$, [8, Theorem 3.3] shows that the trial-sizes Δ^k in that pattern-search method are lower-diminishing, so the bound (2.30) on the subgradients from our Corollary 4 can be used to deduce when a subsequence of the iterates approaches stationarity in this case.

   The pattern-matrices in [8] have the additional property that for any subsequence of pattern-matrices D^k, there exists a positively spanning set $D \subseteq \mathbb{R}^n$ whose elements $d \in D$ are each contained in an infinite number of the D^k. This structure can be exploited to show that for a Lipschitz objective f with compact initial level set $\operatorname{lev}_{f(x^0)} f$, there is an iterate cluster point $\bar{x}$ satisfying the following necessary condition for stationarity at $\bar{x}$:

$$\min_{d \in D} \max_{v \in \partial f(\bar{x})} \langle v, d \rangle \geq 0. \tag{2.34}$$

(e.g., this follows from [1, Theorem 3.7] and [7, Exercise 9.15].) When the objective is smooth $f \in \mathscr{C}^1$, the subgradients reduce to the gradient, and condition (2.34) is also sufficient for stationarity at $\bar{x}$ (since D positively spans $\mathbb{R}^n$).

The next method we will describe is a classic reduce-or-retreat method which we will generalize slightly by including a stopping criterion in terms of a general target-gap. The reason we are interested in this next method in its mostly typical form is because it is notoriously theory-poor, and setting it within our framework reveals avenues for improvement via adaptation of the theory from the other classes of particular methods within the framework.

2.4 Nelder–Mead Method

The classic Nelder–Mead method uses a simplex X^k of $n+1$ n-vectors as the iterate-set but does not identify any element that plays the role of the trial-size Δ^k. Nonetheless, the Nelder–Mead method fits comfortably into our framework (see Lemma 4) if we identify the trial-sizes Δ^k with the volumes vol^k of the iterate-sets X^k:

$$\Delta^k = \overset{k}{\mathrm{vol}} := \frac{\left|\det\left(M^k\right)\right|}{n!}, \tag{2.35}$$

computed in terms of the determinant of the $n \times n$ matrix M^k whose rows come from the "edge-vectors" $x_i^k - x_1^k$ of the simplex:

$$M^k := \begin{bmatrix} \left[x_2^k - x_1^k\right]^{\mathrm{T}} \\ \left[x_3^k - x_1^k\right]^{\mathrm{T}} \\ \cdot \\ \cdot \\ \left[x_{n+1}^k - x_1^k\right]^{\mathrm{T}} \end{bmatrix}. \tag{2.36}$$

A *degenerate* simplex X^k is one with zero-volume (since the corresponding matrix M^k has linearly dependent rows).

The Nelder–Mead trial-set T^k is determined by first arranging the component n-vectors x_i^k of the iterate-set $X^k := \left\{x_1^k, \ldots, x_{n+1}^k\right\}$ from lowest f-value to highest:

$$\left\{x_1^k, \ldots, x_{n+1}^k\right\} \longrightarrow \left\{\breve{x}_1^k, \ldots, \breve{x}_{n+1}^k\right\} \text{ with } f(\breve{x}_i^k) \le f(\breve{x}_{i+1}^k)$$

(with ties broken by the original ordering). The iterate x^k is always the "best" point $\breve{x}_1^k$, and the "worst" point $\breve{x}_{n+1}^k$ is always chosen as the point x_{out}^k to be considered for removal in the reduction test. To define the trial-set, the *centroid* of the simplex face away from the worst point $\breve{x}_{n+1}^k$ is computed:

$$\bar{x}^k := \frac{1}{n} \sum_{i=1}^{n} \breve{x}_i^k,$$

and is then used to compute the following four trial points:

$$\xi_r^k := \bar{x}^k + \theta_{\text{reflect}} \left(\bar{x}^k - \check{x}_{n+1}^k \right) \quad \text{(reflection)}$$

$$\xi_e^k := \bar{x}^k + \theta_{\text{expand}} \left(\xi_r^k - \bar{x}^k \right) \quad \text{(expansion)}$$

$$\xi_o^k := \bar{x}^k + \theta_{\text{contract}} \left(\xi_r^k - \bar{x}^k \right) \quad \text{(outer-contraction)}$$

$$\xi_i^k := \bar{x}^k + \theta_{\text{expand}} \left(\check{x}_{n+1}^k - \bar{x}^k \right) \quad \text{(inner-contraction)} \qquad (2.37)$$

via the fixed Nelder–Mead coefficients of reflection $\theta_{\text{reflect}} = 1$, expansion $\theta_{\text{expand}} = 2$, and contraction $\theta_{\text{contract}} = \frac{1}{2}$ (other choices are possible for these coefficients, but these are standard).

The reduction test in Nelder–Mead is a four-part tree of conditions, and retreat "shrinks" the f-ordered iterate-set toward the iterate x^k by scalar-multiplying $\theta_{\text{shrink}} \in (0,1)$ times each edge-vector $\check{x}_i^k - x^k$ of the simplex. Since the best point $x^k = \check{x}_1^k \in X^k$ is always in new iterate-set X^{k+1} after retreat, the required condition (1.1) that there is no increase in the f-value of the iterate after retreat is satisfied in this case.

2.4.1 Method Outline

Nelder–Mead Method

Step 1. Initialize the (nondegenerate simplex) iterate-set X^0, compute the corresponding initial trial-size $\Delta^0 := \text{vol}^0$ via (2.35), and generate the corresponding trial-set T^0 from (2.37). Fix $\theta_L = \theta_U = (\theta_{\text{shrink}})^n$ and $\theta_{\text{red}} = \frac{1}{2}$, and set $k = 0$.

Step 2. Stopping criterion.

- (Terminate) If the target-gap Γ^k is zero, then

  - Set $\bar{x} = x^k$ as the identified target.
  - Fix all later iterate-sets, trial-sizes, and target-gaps

  $$X^{i+1} = \underbrace{\{\bar{x}, \bar{x}, \ldots, \bar{x}\}}_{m \text{ copies}}, \Delta^{i+1} = 0, \text{ and } \Gamma^{i+1} = 0 \quad \text{for } i \geq k.$$

- Otherwise, continue to Step 3.

(continued)

(continued)

Step 3. Reduce or retreat. Choose $x_{\text{out}}^k := \breve{x}_{n+1}^k$ to be the worst point in the f-ordered iterate-set $\{\breve{x}_1^k, \ldots, \breve{x}_{n+1}^k\}$.

- (Reduce) The reduction test has four cases:

  - (Reflect) If $f(\breve{x}_1^k) \le f(\xi_r^k) < f(\breve{x}_n^k)$ or $f(\xi_r^k) < f(\breve{x}_1^k)$ & $f(\xi_r^k) \le f(\xi_e^k)$, then $\xi^k = \xi_r^k$.
  - (Expand) If $f(\xi_e^k) < f(\xi_r^k) < f(\breve{x}_1^k)$, then $\xi^k = \xi_e^k$.
  - (Outer-contract) If $f(\breve{x}_n^k) \le f(\xi_r^k) < f(\breve{x}_{n+1}^k)$ & $f(\xi_o^k) \le f(\xi_r^k)$, then $\xi^k = \xi_o^k$.
  - (Inner-contract) If $f(\xi_r^k) \ge f(\breve{x}_{n+1}^k)$ and $f(\xi_i^k) < f(\breve{x}_{n+1}^k)$, then $\xi^k = \xi_i^k$.

  If any of the above are satisfied, then:

  - Swap ξ^k for x_{out} in the iterate-set

  $$X^{k+1} = \left\{\xi^k\right\} \cup X^k \setminus \left\{x_{\text{out}}^k\right\},$$

  - Define $\Delta^{k+1} := \text{vol}^{k+1}$ via (2.35);
  - Then continue to Step 4.

- (Retreat) Otherwise, shrink the trial-size $\Delta^{k+1} \in [\theta_L \Delta^k, \theta_U \Delta^k]$, and update the iterate-set by shrinking the f-ordered iterate-set $\{\breve{x}_1^k, \ldots, \breve{x}_{n+1}^k\}$ toward the best point $\breve{x}_1^k$:

  $$X^{k+1} = \left\{\breve{x}_1^k, \breve{x}_1^k + \theta_{\text{shrink}}(\breve{x}_2^k - \breve{x}_1^k), \ldots, \breve{x}_1^k + \theta_{\text{shrink}}(\breve{x}_m^k - \breve{x}_1^k)\right\};$$

  then continue to Step 4.

Step 4. Compute the new target-gap Γ^{k+1}, and generate the new trial-set T^{k+1} from (2.37); increase the iteration-index by 1, and return to Step 2.

The following lemma shows that the Nelder–Mead method is within our framework when we identify trial-sizes with iterate-volumes (2.35).

Lemma 4. *For any iteration-index k with trial-size $\Delta^k = \text{vol}^k$:*

Case (i): If k induces termination, all future trial-sizes $\Delta^{i+1} = 0$ satisfy $\Delta^{i+1} = \text{vol}^{i+1}$.

Case (ii): If k induces a reduction, the new trial-size generated by reduction $\Delta^{k+1} = \text{vol}^{k+1}$ satisfies $\Delta^{k+1} \ge \theta_{\text{red}} \Delta^k$ (as it must to fit the framework).

Case (iii): If k induces a retreat, the new trial-size generated by retreat

$$\Delta^{k+1} \in [\theta_L \Delta^k, \theta_U \Delta^k]$$

satisfies $\Delta^{k+1} = \mathrm{vol}^{k+1}$.

Proof.

Case (i): This follows immediately from the assignment of the iterate-set $X^{i+1} = \{\bar{x}, \bar{x}, \ldots, \bar{x}\}$ upon termination, since $\mathrm{vol}^{i+1} = 0$ for this (degenerate) iterate-set.

Case (ii): According to [5, Lemma 3.1], the four reduction scenarios generate a new iterate-set with volume equal to vol^k scalar multiplied by (respectively) $\theta_{\mathrm{reflect}}$, $\theta_{\mathrm{reflect}} \theta_{\mathrm{expand}}$, $\theta_{\mathrm{reflect}} \theta_{\mathrm{contract}}$, or $\theta_{\mathrm{contract}}$. The result then follows from the definition of $\theta_{\mathrm{red}} = \frac{1}{2}$ which does not exceed any of these scalar multipliers.

Case (iii): According to [5, Lemma 3.1], the new iterate-set X^{k+1} generated by retreat has volume equal to $(\theta_{\mathrm{shrink}})^n$ times the volume vol^k of X^k. The result then follows from the definition of $\theta_L = \theta_U = (\theta_{\mathrm{shrink}})^n$.

□

Remark 17. It follows from the fact that $\mathrm{vol}^0 > 0$ and the proof of Lemma 4 that the trial-sizes $\Delta^k = \mathrm{vol}^k$ in the Nelder–Mead method are always strictly positive prior to termination.

2.4.2 Lower-Diminishing Target-Gaps

Notice that Nelder–Mead is not a descent method since reduction is only guaranteed to reduce the f-value of the iterate in two cases: the second part of the reflect case (when $\xi^k = \xi_r^k$) and the expand case (when $\xi^k = \xi_e^k$). Reflection and expansion also happen to be the only two reduction cases when the trial-size is not decreased, which motivates the following result.

Proposition 12. *For the Nelder–Mead method, if the trial-sizes $\Delta^k = \mathrm{vol}^k$ are lower-diminishing, then any target-gaps Γ^k satisfying*

$$\Gamma^k \geq \gamma > 0 \, \forall k \implies \exists \Delta > 0 \text{ with } f(\check{x}_n^k) > f(\xi_r^k) \text{ whenever } \Delta^k \leq \Delta, \qquad (2.38)$$

are also lower-diminishing.

Proof. This follows from Proposition 2 since the bound $f(\check{x}_n^k) > f(\xi_r^k)$ in condition (2.38) ensures a reduction via either reflection or expansion; and both of these reductions lead to an iterate-set X^{k+1} with volume at least as great as the volume of X^k via [5, Lemma 3.1].

□

Remark 18. Lagarias et al. [5, Lemma 5.2] show that for strictly convex f on $\mathbb{R}^2$ having bounded level sets, the Nelder–Mead iterate-sets always have lower-diminishing volumes vol^k (actually, converging to zero).

2.4.3 Approaching Stationarity

In order to connect target-gaps Γ^k to stationarity, we define the following subgradient-approximating vectors

$$g^k := \left(M^k\right)^{-1} \cdot \begin{bmatrix} f(x_2^k) - f(x^k) \\ f(x_3^k) - f(x^k) \\ \cdot \\ \cdot \\ f(x_{n+1}^k) - f(x^k) \end{bmatrix}, \tag{2.39}$$

in terms of the matrices M^k (2.36) whose rows are the transposed edge-vectors $x_i^k - x^k$ of the simplex. We say that the iterate-sets X^k *avoid simplex degeneracy* if there exists a scalar $\varepsilon > 0$ such that for every iteration-index k prior to termination, the bound holds that

$$\min\left\{\text{von}^k, \inf_{v \in \overline{\nabla} f(x^k)} \text{cosdist}\left(v - g^k, X^k \setminus \{x^k\} - x^k\right)\right\} \geq \varepsilon. \tag{2.40}$$

Here von^k denotes the volume

$$\text{von}^k := \frac{\text{vol}^k}{\max\{\text{rad}^k, \|g^k\|\}^n}$$

of the normalized-simplex

$$\frac{X^k - x^k}{\max\{\text{rad}^k, \|g^k\|\}^n},$$

rad^k denotes the *iterate-radius* of the iterate-set X^k measured from the iterate x^k

$$\text{rad}^k := \max_{x \in X^k} \|x - x^k\|,$$

and $\overline{\nabla} f(x^k)$ is the set of cluster points of gradients evaluated at points of differentiability approaching x^k (see [7, Theorem 9.61]):

$$\overline{\nabla} f(x^k) := \left\{v \,\middle|\, \exists x \to x^k \text{ with } f \text{ differentiable at } x \text{ and } \nabla f(x) \to v\right\}.$$

Fig. 2.1 Iterate-sets
collapsing toward a simplex
with a non-trivial face

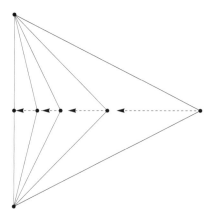

Recall from Eq. (1.11) that the cosine-distance satisfies

$$\mathrm{cosdist}\left(v - g^k, X^k \setminus \{x^k\} - x^k\right) := \sup_{x \in X^k \setminus \{x^k\}} \frac{\langle v - g^k, x - x^k \rangle}{\|v - g^k\| \, \|x - x^k\|}.$$

Notice that under Eq. (2.40), the only way for a subsequence of the iterate-volumes vol^k to approach zero, is to have the iterate-radii rad^k approach zero for the same subsequence. This rules out iterate-sets that collapse toward a degenerate simplex with a nontrivial face as in Fig. 2.1.

Proposition 13. *For the Nelder–Mead method, if the iterate-sets X^k avoid simplex degeneracy (2.40), if the objective function f is Lipschitz continuous on an open convex set O containing the iterates, and if the trial-sizes $\Delta^k = \mathrm{vol}^k$ are lower-diminishing, then the target-gaps $\Gamma^k := \|g^k\|$ are lower-diminishing, and a subsequence of the iterates x^k approaches stationarity.*

Proof. Since the iterate-sets avoid simplex degeneracy (2.40), we know in particular that the iterate-radii rad^k and the target-gaps $\Gamma^k := \|g^k\|$ are simultaneously lower-diminishing with the trial-sizes $\Delta^k = \mathrm{vol}^k$. By assumption, the Δ^k are lower-diminishing, so we focus our attention on the infinite set $\mathcal{K}$ of iteration-indices for which $\mathrm{rad}^k \underset{\mathcal{K}}{\to} 0$ and $\Gamma^k \underset{\mathcal{K}}{\to} 0$. For this subsequence, we know that each element $x_i^k \in X^k$ in the iterate-set can be made arbitrarily close to the iterate x^k:

$$\underset{i}{\mathrm{rad}}^k := \left\| x_i^k - x^k \right\| \underset{\mathcal{K}}{\to} 0 \quad \text{for all } i = 2, \dots, n+1.$$

For any iteration-index $k \in \mathcal{K}$, since f is Lipschitz on O, we know via [7, Theorem 9.61] that $\overline{\nabla} f(x^k)$ is non-empty. So we choose $v^k \in \overline{\nabla} f(x^k)$, which means there is a sequence of points of differentiability $x \to x^k$ with $\nabla f(x) \to v^k$. For any $\eta > 0$, there is an iteration-index K such that $k \in \mathcal{K} \cap [K, \infty)$ ensures that the gradient $\nabla f(x)$ satisfies the bound (e.g., see [7, Eq. 7(16)])

$$\left| f(x_i^k) - f(x) - \left\langle \nabla f(x), x_i^k - x \right\rangle \right| < \frac{\eta}{8} \left\| x_i^k - x \right\|$$

$$\leq \frac{\eta}{8} \left(\left\| x_i^k - x^k \right\| + \left\| x^k - x \right\| \right)$$

$$\leq \frac{\eta}{4} \mathrm{rad}_i^k \qquad (2.41)$$

for all $i = 2, \ldots, n+1$, since both x and x_i^k (by increasing the iteration-index K if necessary) can independently be made arbitrarily close to x^k.

We also always have the following bound:

$$\left| f(x_i^k) - f(x) - \left\langle \nabla f(x), x_i^k - x \right\rangle \right| \geq \left| f(x_i^k) - f(x^k) - \left\langle v^k, x_i^k - x^k \right\rangle \right|$$

$$- \left| f(x^k) - f(x) \right|$$

$$- \left\| \nabla f(x) - v^k \right\| \left\| x_i^k - x \right\|$$

$$- \left\| v^k \right\| \left\| x - x^k \right\|.$$

Since $x \to x^k$, each of the three negative terms on the right side can be assumed to be no less than $-\frac{\eta}{4} \mathrm{rad}_i^k$ for the following three reasons: because of the Lipschitz continuity of f (for the first negative term), because $\nabla f(x) \to v^k$ (for the second negative term), and because v^k is fixed with respect to $x \to x^k$ (for the third negative term). Combining this with our bound (2.41) gives us

$$\left| f(x_i^k) - f(x^k) - \left\langle v^k, x_i^k - x^k \right\rangle \right| \leq \eta \, \mathrm{rad}_i^k. \qquad (2.42)$$

By the definition (2.39) of the subgradient approximating vectors g^k, we know from $M^k \cdot g^k$ that each edge-vector $x_i^k - x^k$ for $i = 2, \ldots, n+1$ satisfies

$$\left\langle x_i^k - x^k, g^k \right\rangle = f(x_i^k) - f(x^k).$$

Substituting this into Eq. (2.42) gives

$$\left| \left\langle v^k - g^k, x_i^k - x^k \right\rangle \right| \leq \eta \, \mathrm{rad}_i^k \quad \Rightarrow \quad \frac{\left\langle v^k - g^k, x_i^k - x^k \right\rangle}{\| v^k - g^k \| \| x_i^k - x^k \|} \leq \frac{\eta}{\| v^k - g^k \|}.$$

Since this holds for any vector $x_i^k \in X^k \setminus \{x^k\}$, we can apply the sufficient cosine-distance component

$$\inf_{v \in \overline{\nabla} f(x^k)} \mathrm{cosdist} \left(v - g^k, X^k \setminus \{x^k\} - x^k \right) \geq \varepsilon$$

of the assumption that the iterate-sets avoid simplex degeneracy (2.40), to conclude that

$$\varepsilon \le \frac{\eta}{\|v^k - g^k\|} \quad \Rightarrow \quad \left\| v^k - g^k \right\| \le \frac{\eta}{\varepsilon}.$$

Since η is arbitrary and ε is fixed, we conclude that $\|v^k - g^k\| \underset{\mathcal{H}}{\to} 0$. Since we have already seen that $\Gamma^k := \|g^k\| \underset{\mathcal{H}}{\to} 0$, we conclude that $v^k \underset{\mathcal{H}}{\to} 0$ as well. The result follows since $v^k \in \partial f(x^k)$ via [7, Theorem 9.61]. $\square$

Remark 19. McKinnon [6] provides an example of a strictly convex smooth objective function on $\mathbb{R}^2$, for which the Nelder–Mead method converges to a nonstationary point in a manner that is visually similar to the illustration in Fig. 2.1. The example in [6] has $\mathrm{vol}^k = \frac{\sqrt{33}}{2^{k+3}}$ and

$$\mathrm{rad}^k = \sqrt{\left(\frac{1+\sqrt{33}}{8}\right)^{2k} + \left(\frac{1-\sqrt{33}}{8}\right)^{2k}},$$

so that the normalized-simplex volume is bounded above as follows:

$$\frac{\sqrt{33}}{2^{k+3}\sqrt{\left(\frac{1+\sqrt{33}}{8}\right)^{2k} + \left(\frac{1-\sqrt{33}}{8}\right)^{2k}}} = \frac{\mathrm{vol}^k}{\mathrm{rad}^k} \ge \mathrm{von}^k.$$

The term on the left approaches zero as k approaches infinity, so the iterate-sets do not avoid simplex degeneracy in this case.

2.5 Incorporating a Degeneracy Check

For each of the four particular examples of reduce-or-retreat methods that we investigated in the preceding sections, conditions ensuring non-degenerate trial-sets T^k were key to establishing that a subsequence of the iterates approaches stationarity. Here is the list of such conditions for comparison.

2.5.1 Non-degeneracy Conditions

• Generalized line-search method: sufficient cosine-distance condition on directions d^k (which determine the trial-sets):

$$\mathrm{cosdist}\left(d^k, \hat{\partial}(-f)(x^k) \cap \mathbb{B}(0; b\|d^k\|)\right) \ge \varepsilon > 0.$$

- Generalized trust-region method: the model functions $\mathrm{mod}^k(p)$ (which determine the trial-sets) have gradients $g^k = \nabla \mathrm{mod}^k(0)$ satisfying:

$$\sup \left\{ \|v - g^k\| \;\middle|\; v \in \bigcup_{x \in \mathbb{B}(x^k;\Delta^k)} \partial f(x) \right\} \leq G\Delta^k.$$

- General pattern-search method: sufficient cosine-distance condition on the pattern-matrices D^k (which determine the trial-sets):

$$\inf_{v \in \partial f(x^k)} \mathrm{cosdist}\left(v, \mathrm{col}\left(D^k\right)\right) \geq \varepsilon > 0.$$

Remark 20. In addition to this non-degeneracy condition, Corollary 4 (showing that a subsequence of general pattern-search iterates approaches stationarity) also assumes the subgradient bound:

$$\inf_{v \in \partial f(x^k)} \left\{ \sup \|\tilde{v} - v\| \;\middle|\; \tilde{v} \in \bigcup_{x \in \mathbb{B}(x^k;\Delta^k)} \partial f(x) \cup -\partial(-f)(x) \right\} \leq G\Delta.$$

This latter bound identifies the kinds of objective functions f to which the corollary applies, but it is not a non-degeneracy condition since it has no connection to the trial-sets T^k.

- Nelder–Mead: avoid simplex degeneracy by bounding normalized-simplex volumes and cosine-distances away from zero:

$$\min\left\{ \mathrm{von}^k, \inf_{v \in \nabla f(x^k)} \mathrm{cosdist}\left(v - g^k, X^k \setminus \{x^k\} - x^k\right) \right\} \geq \varepsilon, \tag{2.43}$$

in terms of the subgradient-approximating vectors

$$g^k := \left(S^k\right)^{-1} \cdot \begin{bmatrix} f(x_2^k) - f(x^k) \\ f(x_3^k) - f(x^k) \\ \cdot \\ \cdot \\ f(x_{n+1}^k) - f(x^k) \end{bmatrix}.$$

In this case, the trial-sets are determined by the vectors in the iterate-set X^k.

Notice that all of these non-degeneracy conditions involve parameters (e.g., ε or G) that are independent of the iteration-index k.

Some reduce-or-retreat methods guarantee non-degeneracy via the structure of the trial-sets used (e.g., the pattern-search method in [8]). However, even when this is not the case, non-degeneracy conditions can be applied a posteriori to determine whether the results of the method can be trusted. Another alternative is to expand the stopping criterion by including a degeneracy check whenever the target-gaps fall below a *gap-threshold* $\gamma > 0$. If degeneracy is detected at this moment (using predetermined values of any parameters in the corresponding non-degeneracy condition), the trial-size Δ^k and/or iterate-set X^k are reset to avoid degeneracy, and the method restarted after shrinking the gap-threshold. Otherwise, the method continues without any changes. In practice, the method can also be stopped when the target-gaps fall below a small enough gap-threshold $\gamma > 0$, since then the target-gaps are "practically zero.

Expanded stopping criterion

Step 2. Stopping criterion.

- (Terminate) If the target-gap Γ^k is zero, then:

    - Set $\bar{x} = x^k$ as the identified target.
    - Fix all later iterate-sets, trial-sizes, and target-gaps

    $$X^{i+1} = \underbrace{\{\bar{x}, \bar{x}, \dots, \bar{x}\}}_{m \text{ copies}}, \Delta^{i+1} = 0, \text{ and } \Gamma^{i+1} = 0 \quad \text{for } i \geq k.$$

- If $\Gamma^k \geq \gamma$, continue to Step 3.
- If $\Gamma^k < \gamma$, check for degeneracy:

    - If the non-degeneracy condition is violated, reset the trial-sizes Δ^k and/or iterate-sets X^k so that the non-degeneracy condition is satisfied, shrink the gap-threshold γ by a fixed factor, and return to Step 2.
    - Otherwise, continue to Step 3.

The modifications necessary to reset when degeneracy is detected are customized to the particular non-degeneracy condition. For instance, when using the non-degeneracy condition (2.43) that we developed for the Nelder–Mead method, a reset would entail altering the iterate-set to increase the normalized-simplex volumes and/or the cosine-distances.

Note that if the target-gaps Γ^k in the augmented reduce-or-retreat framework (using the expanded stopping criterion) are *not* lower-diminishing, then there is a small enough gap-threshold $\gamma > 0$ beyond which the degeneracy check is never again triggered. From that moment forward, the failure of Γ^k to lower-diminish comes entirely from the regular components of the framework (i.e., reduction and retreat). Thus, any failure of the augmented framework to produce lower-diminishing target-gaps Γ^k is essentially due to the same failure within the original framework.

Chapter 3
Scenario Analysis

We analyze seven different scenarios that cover the possibilities for any run of a method within the framework. The convergence properties of the trial-sizes and iterate-sets are explored for each scenario, and the latter analysis is linked to the diameters and radii of the iterate-sets, as well as to the notion of "f-stability" (which entails no increase in the maximum value of the objective on the iterate-set from one iteration to the next).

We have already analyzed the convergence of the iterates x^k in the reduce-or-retreat framework, and we have investigated how an approach to stationarity is connected to the convergence of the target-gaps. Now we will focus on the convergence properties of the other two elements in the framework: the trial-sizes Δ^k and the iterate-sets X^k.

When a reduce-or-retreat method terminates, we completely understand what eventually happens to the trial-sizes and the iterate-sets; and our interest now is to develop our understanding of these elements when the method does not terminate. In order to present the most complete analysis, we consider the reduce-or-retreat framework augmented by the expanded stopping criterion (with degeneracy check), which we give here for convenience.

Augmented Reduce-or-Retreat Minimization Framework

Step 1. Initialize the iterate-set X^0 and the trial-size $\Delta^0 > 0$, and generate the corresponding trial-set T^0. Fix $\theta_L \leq \theta_U$ in $(0,1)$ and $\theta_{\text{red}} \in (0,\infty)$, and set $k = 0$.

Step 2. Stopping criterion.

- (Terminate) If the target-gap Γ^k is zero, then:

  - Set $\bar{x} = x^k$ as the identified target
  - Fix all later iterate-sets, trial-sizes, and target-gaps:

  (continued)

A.B. Levy, *Stationarity and Convergence in Reduce-or-Retreat Minimization*, SpringerBriefs in Optimization, DOI 10.1007/978-1-4614-4642-2_3,

(continued)
$$X^{i+1} = \{\underbrace{\bar{x}, \bar{x}, \dots, \bar{x}}_{m \text{ copies}}\}, \Delta^{i+1} = 0, \text{ and } \Gamma^{i+1} = 0 \quad \text{for } i \geq k.$$

- If $\Gamma^k \geq \gamma$, continue to Step 3.
- If $\Gamma^k < \gamma$, check for degeneracy:

  - If the non-degeneracy condition is violated, reset the trial-sizes Δ^k and/or iterate-sets X^k so that the non-degeneracy condition is satisfied, shrink the gap-threshold γ by a fixed factor, and return to Step 2.
  - Otherwise, continue to Step 3.

Step 3. Reduce or retreat. Choose an outgoing point $x_{\text{out}}^k \in X^k$.

- (Reduce) If the reduction test identifies a satisfactory trial point $\xi^k \in T^k$, then:

  - Swap ξ^k for x_{out} in the iterate-set

  $$X^{k+1} = \left\{\xi^k\right\} \cup X^k \setminus \left\{x_{\text{out}}^k\right\}.$$

  - Choose $\Delta^{k+1} \geq \theta_{\text{red}} \Delta^k$.
  - Then continue to Step 4.

- (Retreat) Otherwise:

  - Update the iterate-set X^{k+1} without increasing the f-value of the iterate (1.1).
  - Shrink the trial-size $\Delta^{k+1} \in [\theta_L \Delta^k, \theta_U \Delta^k]$.
  - Then continue to Step 4.

Step 4. Compute the new target-gap Γ^{k+1}, and generate the new trial-set T^{k+1}; increase the iteration-index by 1, and return to Step 2.

"Scenario I" will refer to the case when a method eventually terminates (at Step 2), and we have already discussed the fact that we completely understand what eventually happens to the trial-sizes and the iterate-sets in this scenario. The other six other possible scenarios for a run of a method can be seen by following the branches to the six terminal nodes on the tree pictured in Fig. 3.1.

For simplicity, we will refer to each of these six scenarios by their terminal node numbers. Thus, "scenario II" refers to a run of a reduce-or-retreat method that leads to a finite number of resets during the degeneracy check, a finite number of

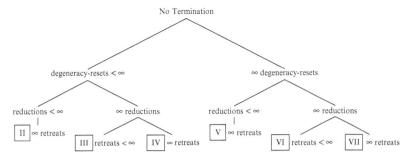

Fig. 3.1 Tree of scenarios when termination is avoided

reductions, and an infinite number of retreats. The *scenario analysis* of a method then refers to the analysis of its convergence properties in a each of the seven scenarios.

For example, since retreat always shrinks the trial-size by at least $\theta_U \in (0,1)$, we can conclude that the trial-sizes Δ^k associated with a reduce-or-retreat method must converge to zero in scenario II. This kind of scenario analysis allows us to provide customized results that may not be revealed by the usual analysis that lumps all of the scenarios together. Our approach thus expands the utility of the convergence analysis because different scenarios may occur for different runs of the same method in practice. Scenario analysis also encourages us to look for conditions guaranteeing a certain scenario for a reduce-or-retreat method since then we can conclude the customized convergence properties for that scenario. For example, [5, Lemma 3.5] proved that only scenario III can occur in the Nelder–Mead method when the objective function f is strictly convex.

3.1 Convergence of Trial-Sizes Δ^k

The key behavior of trial-sizes on which we focus is their convergence to zero, because this often signals when a reduce-or-retreat method is no longer making significant progress.

Theorem 3. *The sequence of trial-sizes converges to zero in any of the following cases:*

Case i: Scenario I.
Case ii: Scenario II.
Case iii: Scenario III, and reduction eventually shrinks the trial-size by at least a fixed factor $\theta \in (0,1)$.
Case iv: Scenario IV, and reduction eventually does not increase the trial-size.
Case v: Scenario V, and degeneracy-resets eventually do not increase the trial-size.

Case vi: *Scenario VI, and together reductions and degeneracy-resets eventually shrink the trial-size by at least a fixed factor $\theta \in (0,1)$.*

Case vii: *Scenario VII, and together reductions and degeneracy-resets eventually do not increase the trial-size.*

Proof.

Case i: The trial-sizes are eventually all zero.

Case ii: There are eventually no degeneracy-resets or reductions, so the result follows since retreat shrinks the trial-size by at least a factor of $\theta_U \in (0,1)$.

Case iii: There are eventually no degeneracy-resets or retreats, so the result follows from the assumption that reduction eventually shrinks the trial-size by at least a fixed factor $\theta \in (0,1)$.

Case iv: There are eventually no degeneracy-resets, so the result follows under the assumption that reduction eventually does not increase the trial-size since retreat shrinks the trial-size by at least a factor of $\theta_U \in (0,1)$.

Case v: There are eventually no reductions, so the result follows from the assumption that degeneracy-resets eventually do not increase the trial-size since retreat shrinks the trial-size by at least a factor of $\theta_U \in (0,1)$.

Case vi: There are eventually no retreats, so the result follows from the assumption that together reductions and degeneracy-resets eventually shrink the trial-size by at least a fixed factor $\theta \in (0,1)$.

Case vii: The result follows from the assumption that together reductions and degeneracy-resets eventually do not increase the trial-size since retreat shrinks the trial-size by at least a factor of $\theta_U \in (0,1)$.

$\square$

Remark 21. The phrase "eventually shrinks the trial-size by at least a fixed factor $\theta \in (0,1)$" includes the situation of eventually having no increase in the trial-size $\Delta^{k+1} \leq \Delta^k$ at any iteration, with an infinite number of iterations mixed in where the trial-size is strictly decreased via $\Delta^{k+1} \leq \theta \Delta^k$.

3.2 Convergence of Iterate-Sets X^k

Any super-vectors $X^k \in \mathbb{R}^{m \cdot n}$ are *convergent* $X^k \to \overline{X}$ if and only if the norm of the difference between X^k and the limiting super-vector $\overline{X} \in \mathbb{R}^{m \cdot n}$ converges to zero:

$$\lim_{k \to \infty} \|X^k - \overline{X}\| = 0.$$

We say that the X^k are *lower-convergent* and write $X^k \rightharpoonup \overline{X}$ if the sequence of difference-vector norms is lower-diminishing:

$$\liminf_{k \to \infty} \|X^k - \overline{X}\| = 0.$$

When the super-vectors represent iterate-sets $X^k = \{x_1^k, \ldots, x_m^k\}$ with m component n-vectors $x_i^k \in \mathbb{R}^n$, and when the limiting super-vector has m identical component n-vectors $\bar{x} \in \mathbb{R}^n$:

$$\overline{X} = \{\underbrace{\bar{x}, \bar{x}, \ldots, \bar{x}}_{m \text{ copies}}\},$$

we say that the X^k are *collapsing* (and write $X^k \to \bar{x}$) when they are convergent and *lower-collapsing* (and write $X^k \rightharpoonup \bar{x}$) when they are lower-convergent. Notice that when $m = 1$ and the iterate-sets are singletons $X^k = \{x^k\}$, the notions of convergence and collapse (as well as lower-convergence and lower-collapse) are equivalent.

Proposition 14. *If an infinite number of the iterate-sets X^k are contained in some bounded set $B \subseteq \mathbb{R}^n$, then $X^k \to \overline{X}$ for some vector $\overline{X} \in \mathbb{R}^{m \cdot n}$.*

Proof. We restrict our attention to the (infinite number of) iteration-indices k for which the iterate-sets $X^k \subseteq B$. For these k, we know that each component n-vector $x_i^k \in X^k$ is also in the bounded set B. Starting with $i = 1$, we conclude that there is at least a subsequence for which the component n-vectors x_1^k converge to some $\bar{x}_1$. Moving to $i = 2$, we know there is at least a sub-subsequence (of the first subsequence) for which the component n-vectors x_2^k converge to some $\bar{x}_2$. Proceeding in this manner successively through all m component n-vectors x_i^k gives the result. □

3.2.1 Iterate-Diameter and Iterate-Radius

One way to study the convergence properties of iterate-sets relies on the concept of *set-diameter* for a (nonempty) finite set $X \subseteq \mathbb{R}^n$:

$$\text{diam}(X) := \max_{x, x' \in X} \|x' - x\|.$$

Notice that $\text{diam}(X) = 0$ if and only if $X = \{\bar{x}, \bar{x}, \ldots, \bar{x}\}$ consists of a finite-number of the same component n-vector $\bar{x} \in \mathbb{R}^n$. The *iterate-diameter* diam^k is then defined to be the set-diameter of the iterate-set:

$$\text{diam}^k := \text{diam}(X^k).$$

We will show below that the lower-collapse of the iterate-sets can be characterized by lower-diminishing iterate-diameters.

A related property that we have already encountered is the *iterate-radius*

$$\text{rad}^k := \max_{x \in X^k} \|x - x^k\|,$$

which measures the greatest distance from the iterate x^k to the rest of the iterate-set X^k. The iterate-radius is zero if and only if there is no "rest of the iterate-set" (i.e., $X^k = \{x^k, x^k, \ldots, x^k\}$). We use the term iterate-radius because rad^k is the radius of the smallest (closed) ball centered at the iterate x^k and containing the entire iterate-set. Clearly, since the iterate x^k is always in the iterate-set X^k, we know that the iterate-radius never exceeds the iterate-diameter

$$\mathrm{rad}^k \leq \mathrm{diam}^k.$$

The next proposition establishes some convergence relationships between iterate-sets, iterate-diameters, and iterate-radii.

Proposition 15. *The following implications always hold:*

(i) $X^k \to \overline{X} \quad \Rightarrow \quad \mathrm{diam}^k \to d < \infty \quad \Rightarrow \quad \mathrm{rad}^k \to r < \infty$
(ii) $X^k \to \overline{X} \quad \Rightarrow \quad \mathrm{diam}^k \to d < \infty \quad \Rightarrow \quad \mathrm{rad}^k \to r < \infty$
(iii) $X^k \to \bar{x} \quad \Rightarrow \quad \mathrm{diam}^k \to 0 \quad \Rightarrow \quad \mathrm{rad}^k \to 0$
(iv) $X^k \to \bar{x} \quad \Rightarrow \quad \mathrm{diam}^k \to 0 \quad \Rightarrow \quad \mathrm{rad}^k \to 0$

Proof. The second implications in all cases follow since the (nonnegative) iterate-radius never exceeds the iterate-diameter; so we focus on the first implications:

(i) Since $X^k \to \overline{X}$, there is a subsequence with the iterate-set X^k converging to some $\overline{X}$. It follows that the corresponding subsequence of iterate-diameters diam^k converges to

$$\mathrm{diam}\left(\overline{X}\right) = \max_{x, x' \in \overline{X}} \|x' - x\| < \infty. \tag{3.1}$$

(ii) Since $X^k \to \overline{X}$, the same argument as in (i) applies to the full sequence.
(iii) Since $X^k \to \bar{x}$, the maximum in Eq. (3.1) is zero since the limit-set $\overline{X}$ consists of m copies of the same n-vector $\bar{x} \in \mathbb{R}^n$.
(iv) Since $X^k \to \bar{x}$, the same argument as in (iii) applies to the full sequence.

$\square$

Remark 22. Lagarias et al. [5, Theorem 5.2] show that the Nelder–Mead iterate-diameters converge to zero for strictly convex objective functions on $\mathbb{R}^2$ having bounded level sets.

If we also know that the iterates are bounded or convergent, most of the implications in Proposition 15 become equivalences.

Proposition 16. *When the sequence of iterates x^k is bounded, the following equivalences hold:*

(i) $X^k \to \overline{X} \quad \Longleftrightarrow \quad \mathrm{diam}^k \to d < \infty \quad \Longleftrightarrow \quad \mathrm{rad}^k \to r < \infty$
(ii) $X^k \to \bar{x} \quad \Longleftrightarrow \quad \mathrm{diam}^k \to 0 \quad \Longleftrightarrow \quad \mathrm{rad}^k \to 0$

When the sequence of iterates x^k is convergent, the following equivalences hold:

(iii) $X^k \to \bar{x} \quad \Longleftrightarrow \quad \mathrm{diam}^k \to 0 \quad \Longleftrightarrow \quad \mathrm{rad}^k \to 0$

Proof. The rightward implications in all cases follow immediately from Proposition 15, so we only need to show that the final property implies the first in each case.

(i) Assuming $\mathrm{rad}^k \to r < \infty$, an infinite number of the iterate-sets X^k satisfy $X^k \subseteq \mathbb{B}\left(x^k; r+1\right)$. Since the iterates x^k are bounded, we conclude that there is a bounded-set B containing all of these iterate-sets X^k; and the result follows from Proposition 14.

(ii) Assuming $\mathrm{rad}^k \to 0$, there is a subsequence with $\mathrm{rad}^k \to 0$. For this subsequence, each component n-vector x_i^k evidently becomes arbitrarily close to x^k. Since the sequence of iterates x^k is bounded, we know it has a cluster point $\bar{x}$; and it follows that there is at least a sub-subsequence for which all of the x_i^k converge to $\bar{x}$.

(iii) Assuming $\mathrm{rad}^k \to 0$, each component n-vector x_i^k becomes arbitrarily close to x^k. Since the sequence of iterates x^k is convergent to some fixed $\bar{x}$, we know that the component n-vectors x_i^k also all converge to $\bar{x}$.

$\square$

Remark 23. The additional assumptions on the iterates x^k are essential to the result in Proposition 16. For instance, any sequence of singleton iterate-sets $X^k = \{x^k\}$ trivially satisfies $\mathrm{rad}^k \to 0$ and $\mathrm{diam}^k \to 0$. However, singleton $X^k = \{x^k\}$ are only lower-collapsing (or collapsing) if the iterates are bounded (or convergent). See Proposition 1 for relatively minor assumptions that ensure bounded iterates.

Remark 24. Proposition 16 does not include an analogue to the pair of implications (ii) from Proposition 15, because those implications cannot necessarily be reversed, even under the stronger additional assumption of convergent iterates. For instance, consider the iterate-sets

$$X^k = \begin{cases} \{0, 1+\frac{1}{k}\} & \text{if } k \text{ even} \\ \{0, -1-\frac{1}{k}\} & \text{if } k \text{ odd} \end{cases}$$

and the iterates $x^k = 0$ (which are trivially convergent) associated with the objective function $f(x) = x^2$. Clearly, X^k is not convergent (though it is lower-convergent to two different limit-sets) even though the iterate-radii and iterate-diameters $\mathrm{rad}^k = \mathrm{diam}^k = 1 + \frac{1}{k}$ converge to 1.

3.2.2 *f*-Stability

The following results use the maximum value $f\,\mathrm{max}^k$ of f on the iterate-set X^k

$$f\,\overset{k}{\mathrm{max}} := \max_{x \in X^k} f(x)$$

to generate level sets

$$\mathrm{lev}_{f\max^k} f := \{x \in \mathbb{R}^n | f(x) \le f\max^k\}.$$

Notice that by the definition of the maximum value $f\max^k$, the iterate-set X^k is always contained in the level set $\mathrm{lev}_{f\max^k} f$. The following definition concerns the relationship between this same level set and the next iterate-set X^{k+1}: the kth iteration of a reduce-or-retreat method is said to be f-stable if $X^{k+1} \subseteq \mathrm{lev}_{f\max^k} f$.

Every iteration inducing a reduction in a reduce-or-retreat method is f-stable. This is because the new iterate $x^{k+1} = \xi^k$ introduced by reduction into the new iterate-set X^{k+1} always has lower f-value than some element of X^k. However, iterations inducing retreat are not necessarily f-stable without additional conditions being met. The following lemma gives one pair of such conditions that apply to the Nelder–Mead method.

Lemma 5. *If a retreat at the kth iteration generates a new iterate-set X^{k+1} whose convex hull is contained in the convex hull of X^k,*

$$\mathrm{con}\left(X^{k+1}\right) \subseteq \mathrm{con}\left(X^k\right), \tag{3.2}$$

and if the convex hull of X^k satisfies

$$\mathrm{con}\left(X^k\right) \subseteq \mathrm{lev}_{f\max^k} f, \tag{3.3}$$

then the iteration is f-stable.

Proof. This follows from the string of containments

$$X^{k+1} \subseteq \mathrm{con}\left(X^{k+1}\right) \subseteq \mathrm{con}\left(X^k\right) \subseteq \mathrm{lev}_{f\max^k} f$$

implied by the two assumptions (3.2) and (3.3). $\square$

Remark 25. The containment (3.3) holds automatically when the level-set $\mathrm{lev}_{f\max^k} f$ is convex since X^k is always contained in $\mathrm{lev}_{f\max^k} f$. Convex objective functions f always have convex level sets [7, Proposition 2.7], as do some non-convex functions (e.g., $f(x) = \sqrt{|x|}$).

Another set of conditions applies to reduce-or-retreat methods like the generalized trust-region method for which retreat might swap in a trial point $\xi^k \in T^k$ for some outgoing point $x_{\mathrm{out}}^k \in X^k$.

Lemma 6. *If a retreat at the kth iteration generates a new iterate-set satisfying*

$$X^{k+1} = \left\{\xi^k\right\} \cup X^k \setminus \left\{x_{\mathrm{out}}^k\right\} \quad \text{with } f(\xi^k) \le f\max^k, \tag{3.4}$$

then the iteration is f-stable.

Proof. The formula in Eq. (3.4) for the iterate-set X^{k+1} and the assumption that $f(\xi^k) \leq f\max^k$ together ensure that $X^{k+1} \in \mathrm{lev}_{f\max^k} f$ since $X^k \subseteq \mathrm{lev}_{f\max^k} f$.   □

One consequence of f-stable iterations is lower-convergent iterate-sets, as long as degeneracy-resets do not expand the level set $\mathrm{lev}_{f\max^k} f$ beyond some fixed bounded set B.

Proposition 17. *Let $\mathscr{D}$ be the set of iteration-indices k at which there is a degeneracy-reset of X^k, and let $\mathscr{D}^*$ be the set of iteration-indices k whose immediate successor is not in $\mathscr{D}$:*

$$\mathscr{D}^* := \{\text{iteration-indices } k \text{ such that } k+1 \notin \mathscr{D}\}.$$

If there exists an iteration-index K for which the level set $\mathrm{lev}_{f\max^k} f$ is contained in a bounded set $B \subseteq \mathbb{R}^n$ and such that:

(a) The iteration-indices $k \in [K, \infty) \cap \mathscr{D}^*$ are f-stable
(b) The level sets for $k \in (K, \infty) \cap \mathscr{D}$ satisfy $\mathrm{lev}_{f\max^k} f \subseteq B$

then, the sequence of iterate-sets is lower-convergent $X^k \rightharpoonup \overline{X}$ to some vector $\overline{X} \in \mathbb{R}^{m \cdot n}$.

Proof. The result follows from Proposition 14 after we show that the iterate-sets X^k are eventually contained in B. We proceed by induction (starting with iteration-index K) to show that

$$\mathrm{lev}_{f\max^k} f \subseteq B \quad \text{for all} \quad k \geq K, \tag{3.5}$$

and the result will follow since we always have $X^k \in \mathrm{lev}_{f\max^k} f$ by the definition of $f\max^k$.

Assuming that $\mathrm{lev}_{f\max^k} f \subseteq B$, we need to establish that

$$\mathrm{lev}_{f\max^{k+1}} f \subseteq B. \tag{3.6}$$

There are two possibilities to consider. First, if $k+1 \notin \mathscr{D}$, then we have $k \in \mathscr{D}^*$, and assumption (a) implies that $X^{k+1} \subseteq \mathrm{lev}_{f\max^k} f$. This in turn implies that $f\max^{k+1} \leq f\max^k$, which means we have the additional containment $\mathrm{lev}_{f\max^{k+1}} f \subseteq \mathrm{lev}_{f\max^k} f$. The desired containment (3.6) then follows from the induction assumption that $\mathrm{lev}_{f\max^k} f \subseteq B$.

The other possibility to consider is $k+1 \in \mathscr{D}$ in which case we get the desired containment (3.6) directly from assumption (b). The condition (3.5) then follows by induction since we assumed the K-case of the containment: $\mathrm{lev}_{f\max^K} f \subseteq B$.   □

Remark 26. The assumption (b) of bounded level sets is a very natural one in the (present) context of minimizing f. Rockafellar and Wets [7, Theorem 1.9] essentially shows that a lower-semicontinuous objective function with bounded level sets will attain its minimum.

Remark 27. For the original reduce-or-retreat framework (without a degeneracy check), the assumptions of Proposition 17 reduce to a bounded level set $\text{lev}_{f\max^K} f \subseteq B$, and the iterations after the Kth one being f-stable.

We now summarize the results of this section to deduce when each of the scenarios associated with a reduce-or-retreat method is certain to generate lower-convergent iterate-sets.

Theorem 4. *The iterate-sets are lower-convergent $X^k \to \overline{X}$ in any of the following cases:*

Case i: Scenario I.

Case ii: Scenario II or IV, and there is an iteration-index K coming after the final degeneracy-reset of the iterate-set with a bounded maximum level set $\text{lev}_{f\max^K} f$, and with retreat for $k \geq K$ generating new iterate-sets X^{k+1} satisfying one of:

$$a. \ \text{con}\left(X^{k+1}\right) \subseteq \text{con}\left(X^k\right) \subseteq \text{lev}_{f\max^k} f.$$

$$b. \ X^{k+1} = \left\{\xi^k\right\} \cup X^k \setminus \left\{x_{\text{out}}^k\right\} \text{ with } f(\xi^k) \leq f\max^k.$$

$$c. \ X^{k+1} = X^k. \tag{3.7}$$

Case iii: Scenario III, and there is an iteration-index K after the final degeneracy-reset of the iterate-set and after the final retreat, with a bounded maximum level set $\text{lev}_{f\max^K} f$.

Case iv: Scenario V or VII, and there is a bounded set $B \subseteq \mathbb{R}^n$ and an iteration-index K such that all iteration-indices $k \geq K$ at which there is a degeneracy-reset of X^k have $\text{lev}_{f\max^k} f \subseteq B$, and retreat for $k \geq K$ generates new iterate-sets X^{k+1} satisfying one of a–c from Eq. (3.7).

Case v: Scenario VI, and there is a bounded set $B \subseteq \mathbb{R}^n$ and an iteration-index K after the final retreat such that all iteration-indices $k \geq K$ at which there is a degeneracy-reset of X^k have $\text{lev}_{f\max^k} f \subseteq B$.

Proof.

Case i: This follows since termination fixes the iterate-set as $\overline{X} = \{\bar{x}, \bar{x}, \ldots, \bar{x}\}$.

Case ii: This follows from Proposition 17 and Lemma 5 (for option a) or Lemma 6 (for option b) since every reduction iteration is f-stable.

Case iii: This follows from Proposition 17 since every reduction iteration is f-stable.

Case iv: This follows from Proposition 17 and Lemma 5 (for option a) or Lemma 6 (for option b) since every reduction iteration is f-stable, and any degeneracy-resets of X^k are assumed to have $\text{lev}_{f\max^k} f \subseteq B$.

Case v: This follows from Proposition 17 since every reduction iteration is f-stable.

$\square$

Remark 28. The generalized line-search method is covered by option c in Case (ii), the generalized trust-region method is covered by options b and c, the general pattern-search method is covered by option c, and Nelder–Mead is covered by option a. Note that for the generalized trust-region method, the trial-set T^k is contained in the ball $\mathbb{B}(x^k; \Delta^k)$, so we have the additional condition $f(\xi^k) \leq f\max^k$ in option b as long as $\mathbb{B}(x^k; \Delta^k)$ is contained in $\operatorname{lev}_{f\max^k} f$ (recall that the iterate x^k is always in this level set).

References

1. Audet, C., Dennis, J.E.: Analysis of generalized pattern searches. SIAM J. Optim. **13**, 889–903 (2002)
2. Ciarlet, P.G., Raviart, P.-A.: General Lagrange and Hermite interpolation in $\mathbb{R}^n$ with applications to finite element methods. Arch. Rational Mech. Anal. **46**, 177–199 (1972)
3. Conn, A.R., Scheinberg, K., Vicente, L.: Introduction to derivative-free optimization. In: MPS-SIAM Optimization series. SIAM, Philadelphia, USA (2008)
4. Kolda, T.G., Lewis, R.M., Torczon, V.: Optimization by direct search: new perspectives on some classical and modern methods. SIAM Rev. **45**, 385–482 (2003)
5. Lagarias, J.C., Reeds, J.A., Wright, M.H., Wright, P.E.: Convergence properties of the Nelder-Mead simplex method in low dimensions. SIAM J. Optim. **9**, 112–147 (1998)
6. McKinnon, K.I.M.: Convergence of the Nelder-Mead simplex method to a nonstationary point. SIAM J. Optim. **9**, 148–158 (1998)
7. Rockafellar, R.T., Wets, R.J.-B.: Variational Analysis. Springer, Berlin (1998)
8. Torczon, V.: On the convergence of pattern search algorithms. SIAM J. Optim. **7**, 1–25 (1997)

A.B. Levy, *Stationarity and Convergence in Reduce-or-Retreat Minimization*,
SpringerBriefs in Optimization, DOI 10.1007/978-1-4614-4642-2,
© Adam B. Levy 2012

Printed by Publishers' Graphics LLC
BT20121002.12.29.62